Veröffentlichungen aus der
Heidelberger Akademie der Wissenschaften

Supplement zu den Sitzungsberichten der
Mathematisch-naturwissenschaftlichen Klasse
Jahrgang 1993/94

F. Vogel R. Grunwald

Patenting of Human Genes and Living Organisms

With Contributions by:

R. G. Adler · F.-K. Beier · A. Dihle
D. Ganten · B. Garthoff · R. Grunwald
Ch. Gugerell · H. Kräußlich · P. Lange
R. Moufang · B. I. Rowland · S. Sahai
A. Saint-Rémy · I. Sela · J. Straus
M. S. Swaminathan · F. Vogel · J. Wagner
B. Yorke · W. Ch. Zimmerli

Springer

Prof. Dr. Dr. h.c. Friedrich Vogel
Institut für Humangenetik
und Anthropologie
Im Neuenheimer Feld 328
69120 Heidelberg, FRG

Dr. Reinhard Grunwald
Deutsches Krebsforschungszentrum (DKFZ)
Im Neuenheimer Feld 280
69120 Heidelberg, FRG

The views expressed in this book do not necessarily represent the decisions of the stated policy of the World Health Organization

---

ISBN 3-540-58148-0 Springer-Verlag Berlin Heidelberg New York
ISBN 0-387-58148-0 Springer-Verlag New York Berlin Heidelberg

Library of Congress Cataloging-in-Publication Data. patenting of human genes and living organisms / F. Vogel, R. Grunwald, (eds.)
p. cm. – (Veröffentlichungen aus der Geomedizinischen Forschungsstelle der Heidelberger Akademie der Wissenschaften) (Supplement zu den Sitzungsberichten der Mathematisch-Naturwissenschaftlichen Klasse; Jahrg. 1994) Includes bibliographical references and index.
ISBN 0-387-58148-0 (acid-free paper: Berlin)
ISBN 3-540-58148-0 (acid-free paper: Berlin)
1. Genomes-patents. 2. Biotechnological microorganisms–Patents. 3. Animals-Patents. I. Vogel, Fredrich, 1925 – . II. Grunwald, R. (Reinhard) III. Series. IV. Series: Supplement zu den Sitzungsberichten der Mathematisch-Naturwissenschaftlichen Klasse; Jahrg. 1994.
K1519.B54P38 1994 346.04'86–dc20 [342.6486] 94-3561 CIP

Printed in Germany

Typesetting: Ulrich Kunkel Textservice, Reichartshausen
SPIN: 10128787 25/3140-543210 - Printed on acid-free paper

Contents

Introduction 1

Session 1

A. Dihle
Begrüßung 5
Welcoming Adress 6

F. Vogel
Patenting of Human Genes and Living Organisms –
A Nightmare for the Scientist 7

J. Straus
Patenting of Human Genes and Living Organisms –
The Legal Situation in Europe 12

R. G. Adler
Remarks on "Technology Transfer and the Genome Program" 30

Discussion: Session 1 35

Session 2

A. Saint-Rémy
Intellectual Property in the Field of Life Sciences 43

Discussion Session 2 52

Session 3

H. Kräußlich
Legal Protection of Living Organisms from the Point of View
of Scientists in Animal Breeding 65

I. Sela
Legal Protection of Living Organisms from the Point of View
of Scientists in Plant Breeding 76

P. Lange
"Patenting" of Living Organisms – Patents and Plant Breeders' Rights –
From the Point of View of Plant Breeders 79

M. S. Swaminathan
Genetic Diversity and the Indian Seed Industry 86

R. Grunwald
Patenting Living Organisms and Their Parts:
The Point of View of Science Administration 94

Ch. Gugerell
Patenting of Human Genes and Living Organisms –
The Current Practice of the European Patent Office 106

Discussion: Session 3 113

Session 4

W. Ch. Zimmerli
Patenting of Human Genes and Living Organisms:
An Ethical Point of View 133

Discussion: Session 4 144

Session 5

S. Sahai
The Patenting of Genes and Living Organisms: The South's View 155

B. I. Rowland
Patenting of Human Genes and Living Organisms:
The Socio-Political Situation in the United States of America 169

Discussion: Session 5 177

Session 6

D. Ganten and J. Wagner
The Human Renin-Angiotensin-System in Transgenic Rats 185

B. Yorke
Which Rules of Patenting are Acceptable
from the Point of View of Industry? 197

F.-K. Beier and R. Moufang
Patentability of Human Genes and Living Organisms:
Principles of a Possible International Understanding 205

B. Garthoff
What does Pharmaceutical Industry expect of Patenting Human Genes
and Living Organisms? 220

Discussion: Session 6 224

Round Table Discussion 232

List of Contributors

Adler, R.G., Dr.
National Institutes of Health
Bethesda, MD 20892, USA

Beier, F-K., Prof. Dr.
Max-Planck-Institut für
ausländisches und internationales
Patent-, Urheber- und Wettbewerbsrecht
Siebertstr. 3, 81675 München, Germany

Dihle, A., Prof. Dr.
Heidelberger Akademie der Wissenschaften
Karlstr. 4, 69117 Heidelberg, Germany

Ganten, D., Prof. Dr.
Max-Delbrück- Centrum für Molekulare Medizin
Robert-Rössle-Str. 10, 13125 Berlin-Buch, Germany

Garthoff, B., Dr.
Bayer AG
Postfach 101709, 42113 Wuppertal, Germany

Grunwald, R., Dr.
Deutsches Krebsforschungszentrum
Im Neuenheimer Feld 280, 69120 Heidelberg, Germany

Gugerell, Ch., Dr.
Europäisches Patentamt
Erhardtstr. 27, 80331 München, Germany

Kräußlich, H., Prof. Dr.
Institut für Tierzucht und Tierhygiene
Veterinärstr. 13, 80539 München, Germany

Lange, P., Dr.
KWS AG
Grimsehlstr. 31, 37574 Einbeck, Germany

Moufang, R., Dr.
Max-Planck-Institut für
ausländisches und internationales
Patent-, Urheber- und Wettbewerbsrecht
Siebertstr. 3, 81675 München, Germany

Rowland, B.I., Dr.
Flehr, Hohbach, Test, Albritton & Herbert
4 Embarcadero Center, Suite 3400
San Francisco, CA 94111-4187, USA

Sahai, S., Dr.
F-31 Green Park (Main)
New Delhi-110016, India

Saint-Rémy, A., Dr.
Commission of the European Communities
12E5 Research, 200 Rue de la Loi, 1049 Brussels, Belgium

Sela, I., Prof. Dr.
The Hebrew University of Jerusalem
The Otto Warburg Center for Biotechnology
P.O. Box 12, Rehovot 76100, Israel

Straus, J., Prof. Dr.
Max-Planck-Institut für
ausländisches und internationales
Patent-, Urheber- und Wettbewerbsrecht
Siebertstr. 3, 81675 München, Germany

Swaminathan, M.S., Prof. Dr.
14 Second Main Road
Kottur Garden, Kotturpuram, Madras 600085, India

Vogel, F., Prof. Dr.
Institut für Humangenetik
und Anthropologie
Im Neuenheimer Feld 328
69120 Heidelberg, Germany

Wagner, J., Dr.
IGBMC
Parc d'Innovation
B.P. 163
67404 Illkirch Cedex, France

Yorke, B.
Leiter der Patentabteilung
Sandoz International AG
Postfach
4002 Basel, Switzerland

Zimmerli, W.-Ch., Prof. Dr.
Universität Bamberg
Lehrstuhl für Philosophie II
96047 Bamberg, Germany

Introduction

Progress in biotechnology in recent decades has created new problems regarding protection of intellectual ownership of results. In what for many basic scientists was regarded as the "golden times" when research in biology, including molecular biology, was a matter of relatively small groups of researchers in universities and other non-profit institutions, this was no real problem. Intellectual ownership was recognized by the scientific community, for example, by adequate citations, but did not, as a rule, have any immediate consequences for the economy or income of the scientist. In recent decades, this has changed completely. With the development of new methods and the extension of our analytical possibilities, results from molecular biological research have been applied in an increasing number of "technical" fields – for example, production of substances useful for diagnostics and therapy of diseases, production of cultural plants such as wheat, corn or rice, and for many other purposes. As a consequence, intellectual property has become a problem with many economic consequences: biological products are being produced and sold no longer only by farmers but by industrial enterprises. Breeders are producing new and much more efficient varieties of seeds for many cultural plants and have to market these products. We are entering a period in history of industrialization in which the "classical" branches such as coal and steel, car production and production of machines, and up to a certain degree, also the chemical industry faces a structural crisis – at least in the highly industrialized countries of the West – and the future of their standard of living will depend more and more on new technologies. These new technologies are information and communication technologies, new materials – and biotechnology. This means that biotechnology is now confronting us with similar problems that had to be faced by other rapidly developing branches, such as production of machine tools or chemicals some 50 or hundred years ago. One of these problems is protection of intellectual and industrial property.

In the course of industrialization, this problem was solved by patent laws. Such laws have existed, for example, in Britain since the 17th century; in most European countries, they were introduced in the second half of the 19th century; the German patent law, for example, was passed in 1877. Patent laws have a double purpose: on the one hand, they protect the industrial property of the inventor; on the other hand, they guarantee that the invention be made available to the general public, because for patenting, publication of the invention is required. In Europe, an invention has to fulfill three requirements in order to be patentable: it has to be new, involve and inventive step ("non-obviousness"), and it has to have commercial utility. Discoveries, on the other hand, cannot be patented. The distinction between discoveries and inventions has given rise to many discussions in recent years that were also reflected by several controversies during this workshop. – Another limitation is that diagnostic or therapeutic methods are not patentable. Exempted from patenting are ultimately inventions not compatible with public order or general principles of morality.

These principles worked quite well in "classical" industries but, the advent of biotechnology has created new problems: some of the "invented" (or discovered?) research results are living beings, for example plants, animals or even parts and products of the human body. Hence, patenting has to be put to the test: are the "classical" principles of patenting sufficient to deal with the new questions? Do they respect morality and the basic rules that regulate life and cooperation in human societies?

Representatives of interested parties such as industrialists and scientists working in the industry argue their case pro patents and are in danger of lacking the necessary sensibility; they should not too readily equate their interests with the interests of society at large. Their views are naturally shared by their lawyers and patent law specialists: patent offices find themselves in a difficult position.

Scientists, on the other hand, and especially medical doctors, often do not really understand what it actually means if something is patented, which rights are involved in the patent, and where its limitations are.

Up to fairly recently, these problems have been discussed in small circles of the immediately interested parties and in patent offices. But in the spring of 1992, two things happened that alarmed the general public and have started a public discussion: first, the European Patent Office in Munich granted a patent of an animal – the so-called "Harvard onco-mouse" – which was patented in the U.S.A. already in 1988 without the European public taking much notive of it. Secondly, the National Institutes of Health (NIH) in the U.S.A. applied for patenting a great number of so-called anonymous cDNA sequences. Meanwhile, most of these applications have been turned down by the U.S. Patenting Office. Meanwhile the new NIH Leadership decided not to appeal the rejections and to withdraw the remaining claims, see Christopher Anderson, NIH Drops Did For Gene Patents, 263 Science 909 (1994); Diane Gershon, U.S. and British Researchers Agree not to Seek Gene Fragment Patents, 367 Nature 583 (1994). Some human gene fragments have been patented in the meantime. At the moment, a great number of patent applications submitted by various institutions have to be decided upon by the European Patenting Office (see the contribution of Gugerell to this workshop). In the face of this situation, the edition agreed to bring together specialists in various fields involved in this issue – scientists, on the one hand, and patent law specialists, representatives of industry, administrators, and politicians, on the other. The Heidelberg Academy of Sciences kindly agreed to host a workshop which took place on July 1 and 2, 1993. The present situation was described by scientists and representatives of industry expressed their opinions on the problems involved. The most important part of the workshop, however, was the continuing discussions between the contributing parties. This discussion was often controversial, and did not lead to a consensus even in the majority of issues, but it helped at least in creating a better mutual understanding.

It is the opinion of the two organizers that a number of principles could help to remove the most obvious (and, at the same time, for the free flow of scientific exchange and public acceptance the most dangerous) stumbling stones:

1) Human genes as parts of the human body as such cannot be patented. They are present in all cells of our bodies, and, therefore, as such cannot be invented but only discovered. But a system involving a gene as well as a procedure for producing a specific gene product, for example, a therapeutically useful functional protein may be patentable provided that it fulfills the existing legal preconditions for patenting (novelty, inventive step, and commercial utility). This implies that other applications for which this gene may be found usable by a later invention are not covered by the patent.

2) Animals normally cannot be patented. Protection of producers' better breeders' rights concerning transgenic animals could be afforded in a similar way, as breeders' rights are being protected at present in plant breeding (see the contribution by Kräußlich). The protection should be limited by exhaustion through, e.g., sale.

3) There was no need to complicate the present regulations for industrial property protection of plant breeders by adding patents on plants. The "classical" breeders' privileges which are balanced by "farmers' privileges" fully serve the purpose of protection, leaving, on the other hand, enough freedom for the farmers. It should be a matter of further discussion whether the possible changes in this system (UPOV) are able to fulfill the farmers' needs, or whether the farmers would fare better with the existing patent rules (exclusion of double protection).

4) A most important point is the protection of freedom of research. All patent laws – and all interpretations of existing laws by patent offices and their counselors – should keep this in mind: any measure that severly inhibits research is wrong and should be fought. It follows that the research privilege that is contained in the European Patent Convention and national patent laws should be interpreted as much in favor of research as ever possible. An important example in this context is clinical research. Research on human hereditary diseases, for example, is performed mainly on patients for whom health insurance systems or private insurances are paying. Still, it is no commercial activity in the normal sense, and should, therefore, enjoy the research privilege. Moreover, the European Patent Convention provides for an 18 month period between patent application and publication. In times of relatively slow industrial development, this regulation might have been adequate. But developments in modern biology are very fast, indeed, requiring a change by which this publication is made much faster.

An important point that has been brought up during this workshop by Dr. Sahai, a participant from India, but has not been discussed thoroughly during the meeting is the influence of patenting in the biological field – especially in plant breeding – on international relationships, and especially on agricultural development and social structure in developing countries. In the opinion of many scientists, plant breeders, and politicians of these countries, acceptance of American and European patents by their legislature would be the consequence of "economic imperialism", leading to severe curtailment of local developments in biotechnology in favor of the superior market power of large Western companies. Further consequences would from their point of view include possible reduction of biological diversity – possibly affecting the long-term food supply of mankind. It was also mentioned that the great majority of cultural plants on which the food supply of all contries is based have originated in areas of the world in which developing countries are situated. Moreover, they have been selected and bred by the farmers of these countries. They are now being utilized by Western industries, and the products are being sold back to these countries. It could be argued that it is completely up to these countries whether they change their own patent laws according to the laws in industrialized countries, and whether they agree to acknowledge foreign patents for their own countries. They should know best the possible advantages and disadvantages. One should take note of the fact, however, that they are now under pressure from international agencies such as the world bank, so their choice is not free.

Obviously, this issue goes far beyond the reach of patent problems. But it is a good example to show, how much the – seemingly very specialized and limited – field of patenting is interwoven with much more general problems. Whether we like it or not – legislative as well as legal practice in patenting will have an important impact on human societies; therefore, it depends, too, on social acceptance within these societies. Here, specialists should not be too confident that their views – only because they appear for themselves to be rational and even benevolent – have to be accepted by society, as well. Scientists, entrepeneurs and patent experts, like other citizens, are living within their respective legal systems and have to comply with their rules. This does not mean, however, that they should regard these rules as optimal. If a workshop like this has any meaning, then, it should give opportunities for scrutinizing these rules and to suggest, if ever possible, improvements.

Heidelberg, fall 1993

Friedrich Vogel　　　　Reinhard Grunwald

Session 1

Begrüßung

Albrecht Dihle

Es ist mir eine große Freude, Sie heute im alten Haus unserer Akademie begrüßen zu dürfen. Es ist, wie Sie sehen, ein schönes und wohnliches Haus, das in den ersten Jahren des 18. Jahrhunderts für eine Familie des regionalen Adels erbaut wurde, später den Badener Großherzögen als Heidelberger Residenz diente und seit dem Jahr 1921 der Sitz der Heidelberger Akademie ist. Ich hoffe und wünsche Ihnen, daß sich die integrative Kraft dieser Räume, die wir bei vielen Gelegenheiten erproben konnten, auch bei Ihrer Zusammenkunft bewähren wird.

Sicherlich gehe ich nicht fehl in der Annahme, daß es angesichts des Themas, das dieser Tagung gestellt ist, ganz besonders darauf ankommt, daß die Teilnehmer in Diskussion und Gespräch zueinander finden. Dabei denke ich nicht nur an die politische Brisanz, die sich aus der hier vorhandenen Verquickung wissenschaftlicher, wirtschaftlicher und nationaler Interessen ergibt. Schon die theoretische Klärung der Situation, in der sich die Genetik befindet, schließt neben den fachwissenschaftlichen auch juristische, ethische, politische und sozioökonomische Fragen ein, und selbst eine Reflexion auf den Sprachgebrauch ist gefragt. Wer hat denn schon die wirklich zulängliche Definiton des Unterschiedes zwischen Entdeckung und Erfindung gefunden?

Daß es Ihnen im Verlauf der kommenden Tage gelingen möge, im vertrauensvollen Austausch Ihrer Gedanken, Einsichten und Erfahrungen auf diesem Felde zu größerer Klarheit und damit auch größerer Handlungsfähigkeit zu kommen, das wünsche ich Ihnen von Herzen, und dazu möge auch unser Haus und die in ihm tätigen helfenden Hände beitragen.

Welcoming Address

Albrecht Dihle

It is a great pleasure for me to welcome you to our Academy house. As you can see, it is a beautiful and comfortable house that was built in the early years of the 18th century for a family of the regional nobility. Later on, it served as the Heidelberg residence of the grand dukes of Baden and since 1921 has been the seat of the Heidelberg Academy. I hope and wish that the power of these rooms to draw people together, which we could test on many occasions, will prove itself also for your meeting.

I am surely not wrong in assuming that in view of the theme set for this conference, it is especially important that the participants in the discussions and conversations reach an understanding with each other. I am thinking not only of the political explosiveness of the combination in this field of scientific, economic and national interests. Already the theoretical clarification of the situation in which genetics finds itself involves not only specialized scientific questions but also legal, ethical, political and socio-economic ones. Even a reflection on language use is required, for who has found a really adequate definition of the difference between discovery and invention?

I wish from my heart that during the coming days of free and open exchange of your thoughts, insights and experiences in this field you will succeed in achieving greater clarity and thus also greater capacity to act. May our house and the helping hands that serve in it also contribute to this.

Patenting of Human Genes and Living Organisms – A Nightmare for the Scientist

F. Vogel

In the spring of 1992, many of us were startled by the news that the National Institutes of Health in the U.S.A. submitted a request for patenting of about 2750 anonymous cDNA sequences of the human genome, which could be parts of genes. Fortunately, the application has been turned down by the U.S. patenting office, but human genes have been patented, and with progress of genetic research in all fields, as well as increasing chances for their commercial exploitation, the run of industrial companies and, more recently, also of research institutes to national and international patenting offices is increasing. But the NIH application has directed public attention to a group of problems that had been discussed in smaller circles for a couple of years. The problems turn out, upon closer scrutiny, to be much more complex than generally assumed. Not only patenting of human genes is at stake, but other living materials have been patented or are under discussion as well – mainly other parts of the human body, – but also animals such as the famous Harvard cancer mouse, – and seeds as results of plant breeding.

Discussion in this field is often characterized by a lack of comprehension of the complexity of these problems. On the one hand, industry points to the big advantages for medical therapy if, for example, human genes are utilized for producing functional proteins. These advantages are undisputed, but these people appear to lack any sensibility for the feelings of a public when it comes to know that certain anonymous institutions have a right on an increasing number of genes present in the bodies of all of us. Such genes can be used not only for the specific purpose which the owner of the patent has in mind, but – with the progress of science – also for possibly many other purposes that cannot be foreseen at present. I will mention one development that has started only recently: this is somatic gene therapy, i.e. introduction of genes into somatic cells for treatment of hereditary diseases. If genes are patented, everyone who wants to treat a patient with somatic therapy has to buy the license from the patent owner. At this point of the discussion, the argument invariably comes that patenting applies only to commercial use of these genes, and that its use for scientific research is – and will be – free. However, this argument deliberately neglects a widely known fact: Research – and, especially, therapeutic research on hereditary diseases – is performed largely on patients. Depending on the health care system, these patients either have to pay the expense themselves or the health insurance is charged. In any case, the hospital uses the genes "commercially." The gene owner is entitled to cash in on it.

Or let me mention another example: New varieties of cultural plants have been produced in the past by "classical" breeding techniques. They could not be patented. However, to make the necessary investments attractive, there were certain breeders' privileges: No one was allowed to simply grow these plants and to sell the seeds. But the farmers were allowed to use them for their own fields and the next harvest. Now, industry pressures legislative bodies very hard to abandon these breeders' privileges in favour of patenting. In the GATT negotiations, for example, developing countries are pressured hard into agreeing with this change. The supposed advantages – also for these countries – are stressed by interested companies and politicians, but no one appears to understand the situation of a farmer in a developing country who is forced to buy his seeds every year anew. For example, strong and logistically perfect police supervision would be necessary to ensure implementation – and who would wish to have such a perfect – and necessarily powerful – police force in countries in which democracy may be in a precarious situation anyway? It should not surprise anyone that representatives of these countries are now accusing companies – and the governments which support their claims – of economic imperialism. The situation is made worse by the present economic recession in the U.S.A. and Western Europe: Biotechnology is regarded – probably correctly – ans one of the most promising future industries that would help to overcome this recession. Conscious of this economic pressure, some industrialists tend to neglect other viewpoints. They lose sight of the fact that an increase of tension and even a political and economic clash between developed and developing countries may lead to a catastrophy for all of us – and first of all for industry itself.

But – of course – industry does not make patenting laws. They are made by the legislative bodies of the various countries. And these legislative bodies are counselled by law specialists. Here, I shall not go into further detail; we are in the fortunate situation that some leading specialists in the field of law, as well as politicians are present at this workshop and will share their experiences and opinions with us. Let me only say so much: Reading earlier publications of law specialists, I was disappointed by an unidimensional and purely formal way of argument; the complexity of the problem was not apparent. At present, this is changing slowly; the European Patent Office in Munich, for example, organized a conference on November 20, 1992, in which achievements of molecular biology and their consequences for patenting were discussed from various points of view.

Unfortunately, the arguments against unlimited patenting, or, to put it somewhat more precisely, – for favour of reasonable limitation of patenting, – were not brought forward at this conference as succinctly as the arguments in favor of patenting. This points to a danger for all those who believe that the future of mankind, – health and happiness of future generations, – depends at least in part on scientific progress – mainly in biology. Any political development that jeopardizes the freedom of science and scientific progress, will be a danger for future generations. Therefore, the present patenting policy of certain sectors of industry, of patenting offices, and of law specialists is a danger for scientific progress, because it strengthens in influential sections of our public an opposition that is directed not only against patenting of human genes and living organisms, but generally against research in molecular biology. The often-repeated argument that, for example, the human genome project intends to enslave human beings by creating the "glass-made man," and that everything is done in the interest of profit-hungry capitalists, – this argument is supported psychologically by patenting of such materials. Those for whom protection of animals is more important than research for improving human health join these ranks utilizing patenting of animals as targets of their attacks. Here, the answer that many of these arguments are irrational is besides the point: Of course, they are, but irrational thinking influences public opinion and political decisions very much, as we have experienced in the past not only in this country. Therefore, everebody interested in scientific progress with as little hindrance as possible by such irrational movements should provide for a moderate patenting policy.

I am grateful to the participants of this meeting – and especially to the invited speakers – that they have accepted our invitation to come here and join in a – probably sometimes controversial – discussion. I hope that the meeting will serve several purposes: First, it will inform us about the actual situation with regard to patenting of biological materials mainly in Europe but also in the United States. Secondly, in bringing together experts from various fields, – science, industry, law and politics, – it may help all of us to listen to the argument of others. In a complex problem field, listening to arguments is only the first, – but a necessary, – step toward mutual understanding. I even hope that such understanding may pave the way for identification of possible solutions for some of the more controversial issues. Such solutions should satisfy two conditions that, on first glance, seem to contradict one another: On the one hand, they should guarantee that groups of people who are interested in the results of biological research do not suffer impediments. I only mention scientists, doctors, but also farmers in developing countries; and there are, of course, many others. On the other hand, the legitimate interests of producers of biotechnological products should be taken into account. Patenting requires that items, in order to be protected by patent, be published and made accessible to all those interested in them. This is, of course, an advantage for science, as well. And finally, – the proceedings of this conference will be published. I am so optimistic as to believe that this publication will have readers, that some of these readers may hold influential positions and that it will help to influence those who have to make political decisions regarding new legislation, but also regarding practical implementation of existing laws, to remain wise and moderate, keeping in mind the long-range interests of human societies. These are also their own long-range interests.

In the following, – the second part of my introduction, – I shall try to sketch some principles that, in my opinion, could help to solve at least some of the controversial problems. I shall go back at least very briefly to some basic principles of patenting that have been established by patenting laws already more than 100 years ago. It is my thesis – which may be disputed by specialists – that a strict adherence to these principles may dispose of at least some of the problems which are bothersome at present.

1. The first condition – which is mentioned very often in recent literature – is that an invention, in order to be patented, should be novel, based on an inventive

activity, and commercially usable. Patentability of certain biological material could possibly be questioned, since this condition may not be fulfilled. Production of transgenic plants, for example, is now becoming a routine procedure, and methods for cloning and isolating genes are applied everyhwere in the world in largely the same way.

2. More important is the second condition: Patenting laws of European countries, including Germany, permit the patentability of inventions. But discoveries are excluded from patenting –even if they can be used later for commercially exploitable inventions. This principle is corroborated by the European patenting law of the 1970ies. X-rays, for example, belong to the discoveries with the widest range of industrial application. But W. C. Roentgen never applied for or received a patent for his discovery nor did the University of Würzburg. And this was in order. Modern discussion tends to neglect this distinction. Strict adherence to it would eliminate many of the problems which we are discussing at present. Human genes, for example, can never be invented, since they are present in all somatic cells of all of us. They can – and should – be discovered. And I doubt very much that a mouse, – and even a transgenic mouse – can be invented. On the other hand, procedures – for example, for isolating genes – can be invented; and they may be patentable. But most work in this field is performed using widely-known standard procedures.

Some specialists may believe that the strict distinction between discoveries and inventions is outdated by biotechnological developments. Here, however, it is necessary to remember that an exclusion of discoveries has a very good reason: Patenting laws, like all other laws, are parts of the value system of a society. They should not contradict this value system. Therefore, it is only logical that the European – and also the German – law excludes inventions from patentability which oppose public order or basic principles of morality. However, freedom of research is an integral part of our value system, as epitomized, for example, by the fact that it is protected explicitly be the German constitution ("Grundgesetz"). Basic scientific research is a noble activity carried out mostly at the expense of the tax payers, and by persons who are privileged socially but receive relatively austere salaries. Results are common property of all mankind, and should therefore be available free of charge. The fathers of patent laws had this in mind when excluding discoveries from patenting. The same applies – and obviously for the same reasons – to methods for surgical or therapeutic treatment of the human or animal body, as well as diagnostic methods that are performed on humans and animals. According to German law, a medical doctor does not have a business; he is not paid in exchange for certain merchandise. Rather, he receives an honorarium as contribution to his livelihood. – Research leading to discoveries belongs to a similar category; therefore, the results cannot be patented.

3. Let us now turn to the other example mentioned at the beginning of my talk, – patenting of cultural plants. Scientists and politicans of developing countries often use the argument that the great majority of cultural plants which we used for our nutrition have been derived from wild plant species in tropical and subtropical areas of our planet, and have been discovered, optimized by breeding

techniques and introduced into human use by farmers in developing countries. Research on history of cultural plants has revealed this beyond any doubt (I only remind you of the work of the great Russian geneticist, Vavilow). These varieties are now available free of charge to companies in developed countries, which, using their patents, want to monopolize sales to those from whom they have received their basic materials. The slogan "getting rich with pilfered genes" exhibits this attitude.

Of course, this argument is wrong from a formal, purely legalistic point of view. Everyone who invents something necessarily starts with the present-day state of the art. But codified law does not necessarily express the over-all value system of the society; in special cases, it may contradict the ethical rules of fairness on which stability of our societies is – or should be – based. In my opinion, it is very well possible that, upon closer scrutiny, patenting of cultural plants may be one of these cases.

I am no law specialist or politician or social scientist. I am a simple medical doctor with some additional training in human and medical genetics. But the problems of patenting biological materials – especially human genes – have been bothering me for quite a while. Therefore, I shall pay close attention to your contributions – the contributions of experts in many different fields. Let us hope that we shall understand one another somewhat better at the end of this meeting and that we may even arrive at some agreement.

Patenting of Human Genes and Living Organisms – The Legal Situation in Europe

J. Straus

I. Introduction

1. It is a challenge indeed for a believer in the positive effects of the patent system on technological, economic and even social progress,[1] to offer a talk immediately after an opening adress calling the patenting of human genes "a nightmare for scientists". Speaking as lawyer in a predominantly scientific environment and being aware of the fact that countless patent applications and specifications are decorated by such foolsounding names of inventors as Anderson, Baxter, Boyer, Cohen, Collins, Galo, Gilbert, Goodman, Leder, Milstein, Powell, Ptashne, Sanger, Schell, Seebourg, Ullrich, Villa-Komaroff, Weissmann, Young, and many, many others, one is automatically reminded of Albert Einstein's theory on relativity: What seems to be a nightmare for some scientists, is seemingly a kind of a dream for some others! Since, however, nightmares as well as dreams occur in the dark of the night, some light has to be brought into this real or apparent contradiction. It certainly would be oversimplistic to try to locate the dividing line between the dream and the nightmare somewhere between the best of the profession and those who try to follow. For in one way or another all scientists are interdependent followers, who stand on the shoulder of their predecessors.

2. What has to be clarified first is, why patents could present for scientists on the one hand a dream and a nightmare on the other, and how would that affect the general public? In order to at least superfluously answer these questions, it has to be understood, what a patent is, what it really does and why we have patents. This temporary deviation from the topic directly assigned to me appears indispensible, for it has to be beyond any doubt that we will all be addressing the same issue.

3. A patent is an exclusive right, limited in time, to exploit an invention, that is to say an instruction for solving a (technical) problem by planned action,[2] provided the solution found is novel, involves an inventive step and is susceptible of industrial application.[3] Third parties may not use the invention without the permission of the patent owner. Where the invention also manifests itself physically, the exclusive rights granted by the patent also extend to the physical form, such

---

[1] Cf. Beier, Significance of the Patent System for Technical, Economic and Social Progress, 11 IIC 563 ss. (1980).

[2] According to the definition offered by the German Federal Supreme Court in the "Red-Dove" decision of March 27, 1969, 1 IIC 136 (1970)

[3] These are patentability requirements of Articles 52–57 of the European Patent Convention (EPC).

as a living organism, be it a microorganism, plant or animal. If the owner of the proprietory right to the physical form is not the same person as the owner of the right to the immaterial good which is the invention and of which the physical form is a manifestation, the holder of each must respect the rights of the other. The granting of a patent does not confer on the patent owner the right to make use of the patented invention! He only may do so if this does not bring him into conflict with the law, such as with health or animal welfare legislation, or with rights of third parties for example. By excluding from the effects of the patent acts done for experimental purposes relating to the subject matter of the patented invention,[4] our legislators have secured that patents by and large cannot inhibit the scientific and technological progress: Experiments aimed also at perfecting, improving and further developing protected inventions, do not infringe the patent.[5]

4. As to the question of "why" we have patents, an exhaustive explanation is neither intended nor possible within this talk. Instead of trying to offer an original new pleading in favour of patents I prefer to refer to two authorities, a recent and a very old one, and to use an empirical example taken out from scientific world, as an object for practical demonstration.

5. Friedrich-Karl Beier, whom we are lucky to have with us today, some ten years ago offered a concise modern interpretation of the widely accepted "patent theories" developed in the fifties by Fritz Machlup. Professor Beier's answer is the shortest possible and exactly to the point. We grant patents:

- to recognise the intellectual property of the inventor;
- to reward the inventor for his useful services;
- to encourage inventors and industry to invent, invest and innovate; and
- to further the early disclosure and wide dissemination of technical knowledge.[6]

It goes without saying that in market economies the inventor can receive his reward only via the market, i.e. via successful commercialisation of his/her invention. The patent does only secure a privileged market opportunity.

6. For lawyers a patent has always represented a bargain between the inventor and the society: In exchange for the disclosure of his/her invention, the inventor is granted an exclusive right of exploitation, What judge Buller, the old authority, stated in the 1785 famous Arkwright case in this context is still valid and decisive and therefore should be recalled:

---

4 See for details on the experimental use issue Chrocziel, Die Benutzung patentierter Erfindungen zu Versuchs- und Forschungszwecken, Cologne, Berlin, Bonn, Munich 1986.

5 In this context, however, it is admitted that even the modern European law maker did not sufficiently clarify all essential issues at stake. At present the tendency of case law to interprete the respective provisions extremely narrowly is observed. See e.g. Straus, Zur Zulässigkeit klinischer Untersuchungen am Gegenstand abhängiger Verbesserungserfindungen, 1993 GRUR 308 ss.

6 11 IIC 574 (1980).

"It is clearly settled at law that a man, to entitle himself to the benefit of a patent for a monopoly must disclose his secret and specifiy his invention in such a way that others may be taught by it to do the thing for which the patent is granted, for the end and meaning of the specification is to teach the public after the term for which the patent is granted, what the art is, and it must put the public in possession of the secret in as ample and beneficial way as the patentee himself uses it . . . , for the patent is the reward which, under the Act of Parliament, is held out for a discovery; and, therefore, unless the discovery be true and fair the patent is void."[7]

7. Thus, the *quid pro* quo for granting the patent is the disclosure of the invention, of the solution found to the problem for which protection is sought. As in the time of Judge Buller, this disclosure must be sufficiently clear for the invention to be reproduced by one skilled in the art. Taking into account the specific problems posed to inventors in their attempts to repeatably describe inventions in the field of living matter, our patent laws have adjusted the disclosure requirements: If the written description is not sufficient to reproduce the invention, the inventor may and must furnish further information by depositing biological material in depository institutions which are publicly accessible, longterm and impartial, so that, within 18 months of the first filing date at latest, e.g. European patent law guarantees transparency. From that point onward the content of a patent application, and also any deposited material, are available to the public, and experiments by third parties aimed at improving and perfecting the invention are both possible and permitted in patent law terms.[8] In this context it should also be added that while it is certainly true that expectation of profits discourages open discussions of technical detail during the critical phase before patent filling,[9] it is also true that the situation is worse if scientists producing technically and commercially applicable research results have no prospects for patent protection. Then secrecy is the only alternative and in most cases also the alternative, which is actually used.[10]

---

[7] 1 WPC 60 (64) (1785), quoted according to Wallace and Williamson, The Law and Practice Relating to Letters Patents for Inventions, London 1990, p. 187.

[8] See on problems of deposit and release of biological material Straus and Moufang, Deposit and Release of Biological Material for the Purposes of Patent Procedure – Industrial and Tangible Property Issues, Baden-Baden 1990. See also Crespi, The Micro-Organism Deposit System in European Patent Law – An Appraisal of Current Proposals, 24 IIC 1 ss. (1993).

[9] See Kevles and Hood, Reflections, in: Kevles and Hood (eds.), The Code of Codes, Cambridge, Mass., and London 1992, pp. 300 ss. (at 312).

[10] This is not just a "streamlined" position of the patent system promoters, but the view expressed by prominent scientists themselves. See the clear statement made by Professor Brenner in a discussion held on problems of genome exploitation. It reads: "The argument is that if you didn't have patenting it would have been kept a secret." Science, Law and Ethics, Chichester, New York, Brisbane, Toronto, Singapore 1990, p. 145.

8. Since we will have a number of speakers, who will adress the very topic of this workshop under the point of view of scientists actively involved in the field, I will resist the temptation to try do deduce from my brief remarks on some fundamentals of patent law any general perception as to whether patents may be a good cause for nightmares or for dreams of scientists. Instead I would remind you only of the experience linked to the discovery of cephalosporins and their resulting use in medicine: In 1948 Giuseppe Brotzu, a professor of bacteriology at University of Cagliari, Sardinia, isolated an antibiotic substance produced by a fungus, a strain of cephalosporium acremonium, which he found in the sea near a sewage outfall from the city of Cagliari. Because of his very limited financial means, he bypassed animal tests and administered the material to human patients with, e.g. typhoid fever, who showed some improvement. After failing to interest any Italian pharmaceutical company in his findings, he published them in a one-issue journal of his Institute and, more importantly, reported them via a former British public health officer in Sardinia to Sir Edward Abraham of the University of Oxford in England. The rest of the success story should be known. Cephalosporins from Oxford, protected by patents, have secured the financial basis for the British National Research and Development Corporation (NRDC, now British Technology Group – BTG) and Robert C. Woodward from Harvard, who synthesized cephalosporin C, received the Nobel Prize in 1966. Now, one might ask, where is the gist of the story? In the context of interest, it has to be seen in the fact, that Brotzu's original dream of saving ill patients by a new drug and to possibly raise some badly needed funds for his ailing institute, blighted due to the lack of interest by the Italian pharmaceutical industry, which in turn was due to the then non-existent patent protection for pharmaceuticals in Italy. What started as a dream ended as a nightmare not because of patents but because of lack of patents. There is one additional lesson to be learned from this experience: It is not enough to have good scientists, who discover new strains of microorganisms and even isolate new products produced by such organisms. In order to get a product on the market, additional investments are needed, which require adequate protection. Eventually, the case of cephalosporin, as many other cases do, indicates that patents on substances did not hinder the subsequent development in the field. Nearly 20 cephalosporins are available today and many more are on the synthetic horizon.[11]

9. I hope very much that all of you will forgive me this quite considerable excursion into some patent law aspects of more general interest for scientists. As the first invited speaker of the workshop I simply felt the need to address the issue of patenting in a more general context, not least to show that in the past many scientists successfully made use of patents to their and the general public's benefit, and without improperly inhibiting the progress of science. The best and most relevant example, which I have not yet mentioned, is the Cohen and Boyer US-patent of Stanford and UC San Francisco. This patent covers the basic technology

[11] Cf. Roberts, Serendipity – Accidental Discoveries in Science, New York, Chichester, Brisbane, Toronto and Singapore 1989, pp. 204 ss.

of inserting foreign genetic material into bacterial plasmids. Last year it yielded almost US $ 15 million and will have earned by the date of its expiry in 1997 some US $ 90 million.[12] Had that patent seriously inhibited scientific or technological developments, or otherwise posed insurmountable problems to the scientific community, this workshop would have probably been devoted to other issues. Let me now turn to the topic actually assigned to me, namely the European legal situation with respect to patenting human genes and living organisms.

II. Patenting of Living Organisms – Brief Historical Retrospect

a) General

10. While it goes without saying that patenting of genes is a very recent issue, inseparable from rDNA techniques, patenting of living organisms, contrary to some widely disseminated views, is not. The first known patent for a living organism, a yeast, was granted in Finland exactly 150 years ago![13] Without going into any details of the respective developments, one has to observe, e.g. for Germany, that patents were granted from the very beginning, i.e. from 1877 onward for culturing of yeasts, the preparation of bread and beer, and the production of vinegar. Later, patents were granted for the fermentative production of butyl alcohol and acetone with the aid of bacteria. After the discovery of the antibiotic penicillin, patents were also granted for auroeomycin, streptomycin, tetracyclin, etc., produced with the aid of known or yet to be discovered microorganisms through fermentative methodes.[14] In the thirties the German Patent Office started to allow claims for agricultural cultivating processes, as well as for new varieties of plants.[15] One has to note that since 1891 the effects of process patents according to the German law have been extended to their direct products, too.

11. It was before the advent of genetic engineering and more than 20 years before the "Onco-mouse" decision of the Technical Board of Appeal of the European Patent Office (EPO) that our Federal Supreme Court in the famous and at that time undisputed "Red Dove" decision of 1969 explicitly declared animal breeding methods and their resulting "products" patentable subject matter.[16] When referring

---

12 See the report by Sally Lehmann, Stanford Seeks Life after Cohen-Boyer Patent Expires, 363 Nature 574 (17 June 1993).

13 Patent No. 3 for "Ferment de Poche" of 1843. See Lommi, Problems Related to Protection of Plants, 1989 NIR 1.

14 See on these developments Beier, Crespi and Straus, Biotechnology and Patent Protection – An International Review, OECD Paris 1985, pp. 22 ss. with further references.

15 See for details Straus, Der Schutz biologischer Erfindungen, insbesondere von Pflanzenzüchtungen, in: Beier, Kraft, Schricker and Wadle (eds.), Gewerblicher Rechtsschutz und Urheberrecht in Deutschland, Festschrift zum hundertjährigen Bestehen der deutschen Vereinigung für gewerblichen Rechtsschutz und Urheberrecht und ihrer Zeitschrift, Weinheim 1991, pp. 363 ss. (382 ss.).

16 1 IIC 136 (1970).

to the concept of invention to be applied in the field of biology, the Court stated inter alia:

> "We are concerned with the question whether biological phenomena and forces can be treated the same way as those of a technological nature. In this connection it is again of little importance what the legislator in 1877 considered to be "technology", but rather how the biological phenomena and forces are to be understood and classified in the present state of sciences . . . A teaching to methodically utilise controllable natural forces to achieve a causal, perceivable result, could be considered patentable, provided that teaching meets the general prerequisites of industrial application, novelty, etc."[17]

b) Current Statutory Situation and its Roots

12. The concept of patentable invention currently applied under the European Patent Convention (EPC) and the national laws of most European countries goes back to the Strasbourg Convention on the Unification of Certain Points of Substantive Law On Patents For Invention of 1963. Under the general and mandatory rule of Article 1 of this Convention, the Contracting States were obliged to grant patents "for any inventions which are susceptible of industrial application, which are new and which involve an inventive step." For two widely differing groups of inventions, however, the Contracting States were offered by Article 2 a free hand to provide or not to provide for the grant of patents.

13. the first group is of very general character and relates to inventions, the publication or exploitation of which would be contrary to "ordre public" or morality, but with the explicit limitation that the exploitation shall not be deemed to be so contrary merely because it is prohibited by a law or regulation. The second group relates to inventions of plant or animal varieties or essentially biological processes for the production of plants or animals. Here again, an important limitation of the freedom of Contracting states has been provided for: They have to grant patents for microbiological processes and the products thereof.[18]

Whereas the first group of allowed exceptions was modelled by and large along the lines of similar provisions already existing in the national patent laws, the second one reflected some pecularities, which were predominantly linked to patenting in the kingdom of plants. Not withstanding the fact that patents had been granted for new varieties of plants in a number of European countries[19] patenting in this field has encountered problems especially in respect to the repeati-

---

17 1 IIC 138 (1970).

18 See for more details Beier, Crespi and Straus, op. cit., footnote 14, pp. 28 ss.; Beier and Straus, Genetic Engineering and Industrial Property, 1986 Industrial Property 447 s.

19 Cf., e.g. Straus, Industrial Property Protection of Biotechnological Inventions – Analysis of Certain Basis Issues, WIPO Doc. BIG/281, Geneva 1985, para. 50 ss. at pp. 60 ss.

bility requirement and, more importantly, was unpopular with the powerful agricultural administrations.[20] Under the decisive influence of the latter, a new special system of exclusive rights for protecting new varieties of plants has been estabilshed, when in Paris the International Convention for the Protection of New Varieties of Plants (UPOV-Convention) was signed in 1961.[21] The drafters of Strasbourg viewed the UPOV-Convention as an appropriate alternative for protecting plant varieties, therefore they refrained from imposing respective obligations on the Contracting States in Patent Laws.[22]

14. The reason for the parallel treatment of inventions in animal varieties are less clear, since a special system for their protection neither existed nor was it under consideration. The prevailing view seemingly was that animal varieties and animal breeding processes would not fit well into the classical forms of patent protection.[23] Moreover, the then state of the art in animal breeding was considerably lagging behind that in the field of plants, so that there was virtually no pressure on the law makers. The parallel treatment of plant and animal varieties and the indirect link to the UPOV-Convention revealed, however, that a distinction was made between plant or animal varieties as specific groupings of plants or animals which the contracting states were allowed to exclude from patent protection and plants or animals as such. This is also demonstrated by the use of the terms, plants or animals in respect to essentially biological processes for their production.[24]

15. What was left at the discretion of the Contracting States under the Strasbourg Convention, eventually led to be strict exclusionary provisions of Article 53 (a) and (b) of the EPC and the pertinent national patent law provisions. This metamorphosis has taken place without any substantial reconsideration against the ongoing scientific developments in the field of molecular biology and genetics. The resulting situation therefore is viewed by the original creators of the EPC as

---

20 Cf. Bent, Schwaab, Conlin, Jeffrey, Intellectual Property Rights in Biotechnology Worldwide, New York 1987, pp. 46 ss.; Neumeier, Sortenschutz und/oder Patentschutz für Pflanzenzüchtungen, Cologne, Berlin, Bonn, Munich 1990, pp. 21 ss.

21 For the establishment and the content of the 1961 Act of this Convention, cf. Bent, Schwaab, Conlin and Jeffrey, op. cit., footnote 20, pp. 55 ss.; Neumeier, op. cit., footnote 20, pp. 33

22 Cf., e.g., Moufang, in: Europäisches Patentübereinkommen – Münchener Gemeinschaftskommentar (eds. Beier, Haertel und Schricker), Cologne, Berlin, Bonn, Munich 1984 ss., delivery No. 15, Art. 53 Marginal Notes 12 ss.

23 Cf. Pfanner, Vereinheitlichung des materiellen Patentrechts im Rahmen des Europarates, 1962 GRUR Int. 545 ss. (at 547); and Straus, Ethische, rechtliche und wirtschaftliche Probleme des Patent- und Sortenschutzes für die biotechnologische Tierzüchtung und Tierproduktion, 1990 GRUR Int. 913 ss. (at 920).

24 This has been correctly pointed out in the decision of the Technical Board of Appeal of the European Patent Office (TBA) of October 3, 1990, 222 IIC 74 ss. (No. 4.6 at p. 81) (1991) – "Onco-Mouse II".

an unfortunate one.[25] In consequence, the statutory basis for patenting (human) genes and living organisms in Europe presents itself as follows:

(i) According to the general rule of Article 52 (1) EPC, European patents shall be granted to any inventions which are susceptible of industrial application, which are new and which involve an inventive step; however, discoveries, if claimed as such are not regarded as inventions (Article 52 [2] and [3]);

(ii) for inventions the publication or exploitation of which would be contrary to "ordre public" or morality, patents cannot be granted, provided that the exploitation shall not be deemed to be so contrary merely because it is prohibited by law or regulation in some or all of the contracting states (Article 53 [a]);

(iii) European patents also cannot be granted for plant or animal varieties or essentially biological processes for the production of plants or animals, but ought to be granted for microbiological processes and the products thereof (Article 53 [b]).

How these provisions have influenced the patent granting practice, especially of the European Patent Office, should be examined in brief in the context of the respective groups of inventions to be addressed.

III. Patent Granting Practice of Interest

a) (Human) Genes

16. You will have noticed that the EPC is silent as far as genes are concerned. The same is true for national patent laws. Having regard to the fact that the fundamental physical and functional unit of heredity, *the gene,* is an ordered sequence of nucleotides located in a particular position on a particular chromosome, which encodes a specific functional product, such as a protein or RNA molecule,[26] this may not come as a surprise. Deoxiribonucleic acid, the molecule that encodes genetic information, is a biochemical substance and has been treated by patent offices in the same way as other chemical substances. Thus, patents have been granted for a considerable number of genes encoding various human proteins, e.g. for various interferons,[27] granulocyte colony stimulating factors (GCSF),[28] granulocyte macrophage colony stimulating factors (GmCSF), growth hormones, blood coagulating factors, erythropoietin[29] and the like. No differen-

---

25 See, e.g., Armitage, Updading the European Patent Convention, 1990 GRUR Int. 662 ss.; Savignon, Considérations sur les principes du rapprochement des législations en matières de brevets: La brevetabilité des êtres vivants, 1990 GRUR Int. 766 ss.

26 According to the definition offered in: Kevles and Hood, op. cit., footnote 9, p. 379.

27 E.g., the European Patent (EP) No. 32 134 for the IFN αtype. Cf. also the decision of TBA of February 16, 1989, 1990 O. J. EPO 335 – Alpha-interferons/BIOGEN.

28 EP No. 0237545 (Amgen); EP No. 0215126 (Chugai).

29 EP No. 148605 (Amgen).

tiation has been made whether the respective DNA sequence coded for a human or, e.g. a monkey protein. In some patents, as for instance in the European patent issued for erythropoietin, claims were allowed for a specific DNA sequence encoding human erythropoietin as well as for a specific DNA sequence encoding a monkey species erythropoietin. Moreover, the claims granted covered both the genomic and the copy DNA (cDNA), i.e. the DNA "freed" of noncoding regions, the so-called introns.[30]

17. Since in all referenced cases the complete DNA sequence of the respective genes has been disclosed and the protein for which the gene is coding, as well as its useful properties, indicated, the main problems to be overcome by applicants were those of the novelty and inventive step requirement.[31] The fact that discoveries as such are not regarded inventions under the EPC, however, did not pose substantial problems: in the case of cDNA sequences, this problem does not exist at all, since they do not occur in nature. But also as far as genomic DNA sequences are concerned, the prevailing view and the patent office practice is that human technical intervention is required to recognise them and to produce them in a reproducable manner. The subject matter of such application, therefore is, an invention and not a mere discovery.[32]

18. How the European Patent Office will resolve the problem of the very much discussed and disputed applications, which relate to DNA stretches with functions which are not yet known, as in the case of the application filed by the National Institutes of Health (NIH – Inventor Greg Venter et al.) in June 1991 and published on 7 January 1993 as PCT application,[33] is not simple to predict. Whereas the US-PTO in its preliminary decision of December 1992 could reject most of the claims on the basis of the lack of utility or novelty,[34] the EPO, even if of the same opinion, will have to reason differently. The requirement of "industrial applica-

30 Claim 2 of the EP No. 148605 reads as follows: A DNA sequence according to Claim 1 encoding human erythropoietin; Claim 3, A cDNA sequence according to Claim 1 or 2. Claim 4 is directed to a monkey species erythropoietin coding DNA sequence according to Claim 3, and, e.g., Claim 6, a genomic DNA Sequence according to Claim 1 or 2.

31 For a good selection of decisions of the TBA of the EPO on novelty and inventive step issues of biotechnological inventions see Jaenichen, The European Patent Office's Case Law on the Patentability of Biotechnology Inventions, Cologne, Berlin, Bonn, Munich 1993, pp. 95 ss., 109 ss.

32 Cf. the EPO Examination Guidelines Part C-IV, 2.1

33 International Publication No. WO 93/00353. As reaction to the NIH patent activity, similar applications have been filed by the British Medical Research Council (MRC), too. Cf. Roberts, MRC Denies Blocking Access to Genome Data, 256 Science 1583 (1991).

34 The Primary Examiner rejected claims 1–24 under 35 USC § 101 because of the lack of patentable utility as well as under 35 USC § 102 (b) – in view of anticipation by Connor et al. In respect to the first argument the examiner stated that "the mere mention of general possible uses is not sufficient to establish a definite utility . . . Given what is disclosed in the instant application it would be necessary for one to do further work in order to establish a utility for any of the nuclear types embraced by the claims" (p. 6 of the examiner's notice of December 2, 1992).

tion" as defined in Article 57 EPC, which is often considered to be the European counterpart of the "utility" requirement under US patent law, can hardly be invoked, since it has to be regarded as fulfilled, if an invention can be made or used in any kind of industry, including agriculture. It is beyond doubt that the DNA stretches at stake can be made in industry. In Europe, therefore a different approach is needed. The prerequisite of inventive step (Article 56 EPC) might be an appropriate choice. In a workshop organized in November 1992 by the Human Genome Organization (HUGO) Europe, agreement has been widely achieved among participants that the cDNA strategy used by Greg Venter and his group, was routine and obvious and might soon lead to a situation in which such invention will not be able to meet the criterion of inventiveness.[35] Moreover, it has to be borne in mind that within the next two years, such a high proportion of the 100 000 or so genes in the human genome will be represented by fragmentary cDNA sequences in the public domain, that the prospects to get patents granted even for full length cDNA sequences as such will be rapidly decreasing.[36]

19. If there is a serious problem related to patenting of genes, that problem most probably has to be seen in the far reaching effects of the broad product protection accorded especially to pioneer inventions from the end of the seventies and beginning of the eighties. Since the pertinent inventions were based on a relatively modest and incomplete understanding of the functional correlations, e.g. between the gene constructs and vectors used in transformation or transfection methods and the host organisms used for expression, but were granted product protection for DNA sequences *per se,* they may constitute considerable obstacles for the exploitation of later inventions based on advanced knowledge of such correlations and providing for essentially improved, e.g. pharmaceutical substances which are biologically more active and with fewer side effects. Here possibly the law maker will have to provide for statutory relieves.[37]

b) Living Organisms

20. As has been already mentioned, different categories of living organisms with respect to their patentability are offered different treatment under the EPC. On the one hand it was fairly clear from the outset that microorganisms as products of microbiological processes are to be offered product protection. On the other hand, the exclusion of plant or animal varieties and essentially biological processes for the production of plants or animals was also beyond doubt. With respect to the latter, however, it was not clear, what exactly has to be understood under the

---

[35] The results of the HUGO-workshop are reproduced in a report sent to the EC Commission on 24 June 1993.

[36] This was concluded by a EC Workshop on cDNA Strategies, held in September 1993 in Paris (according to the HUGO-Workshop Report, op. cit., footnote 35, under 3.6).

[37] See on pertinent problems Straus, 1993 GRUR 309, 310; Bicker, as quoted by Hoffmann, Patente dürfen die Markteinführung nicht blockieren, Handelsblatt of March 12, 1993; Collard, Limited Patent Protection for Proteins, 10 Canadian Intellectual Property Review 25 ss. (September 1993).

terms plant or animal variety and where to draw the dividing line between inventions in plant or animal varieties, which are excluded from patent protection and other inventions in plants and animals. Since with respect to microorganisms the only severe problem, namely the acceptance of deposits of strains in order to meet the sufficient disclosure requirement for product protection, too, has been resolved at an early stage of development of the European patent system,[38] there is no need for any special discussion on patenting of microorganisms in the context of this contribution.

21. Corresponding to the developments in natural sciences, the EPO was first confronted with patent applications in which protection was sought for inventions in plants. Claims in the respective applications were directed to plants, plant cells or plant tissue, modified by the use of either rDNA or somatic cell hybridization techniques. For long these applications have been predominantly questioned under the Article 53 (b) provision. The question, which the EPO had to answer therefore was whether the respective claims were directed to plant varieties and whether protection to be granted could extend to plant varieties, irrespective of the concrete wording of the claims. The principles still valid with respect to the granting patents for genetic inventions in plants or animals have been established by a decision of the Technical Board of Appeal of 26 July 1983.[39] The Technical Board of Appeal has first clarified that no general exclusion of inventions in the sphere of animate nature can be inferred from the EPC. For the interpretation of the terms "plant variety", the Board referred to the definition originally used in the UPOV Conventions 1961 Act. It stated that Article 53 (b) EPC prohibits only the patenting of plants or their propagating material in the genetically fixed form of the plant variety, i.e. a multiplicity of plants which are largely the same in their characteristics and remain the same within specific tolerance after every propagation or every propagation cycle, and that the very wording of that provision precludes the equation of plants and plant varieties which would also be at variance with the general sense of the provision.[40] The EPO has upheld this questionable nexus between the term plant variety used in the EPC and in the UPOV Convention, although the 1978 UPOV Act did not contain any definition and the definition introduced in the 1991 Act (not yet in force) has introduced new elements differing quite considerably from those forming part of the definition under the 1961 Act.[41] In an as yet unpublished decision of 5 June 1992[42] in which

38 By decision of the President of the EPO of December 11, 1981, Part C-IV, 3.5 of the Guidelines for Examination in the EPO has been amended accordingly, 1982 O.J. EPO 19.

39 T49/83, 1984 O.J. EPO 112 – "Propagating material/CIBA-GEIGY".

40 No. 2–4 of the reasons for the decision, loc. cit., pp. 114 ss.

41 For the new definition see, e.g., Greengrass, The 1991 Act of the UPOV Convention, 1991 EIPR 466 sds. (467); Teschemacher, Die Schnittstelle zwischen Patent- und Sortenschutzrecht nach der Revision des UPOV-Übereinkommens von 1991, in: Bruchhausen et al. (eds.), Festschrift Rudolf Nirk zum 70. Geburtstag, Munich 1992, pp. 1005 ss. (1010 ss.); Straus, Pflanzenpatente und Sortenschutz – Friedliche Koexistenz, 1993 GRUR 794 ss.

claims were directed not only to DNA shuttle vectors, but also to a plant cell, and to a plant or plant tissue grown from that plant cell, the Opposition Division has confirmed the patent granting policy developed on the basis of CIBA-GEIGY decision also for the time to come after the new UPOV Act in which the present ban on double protection has been removed, has entered into force. Since the claims at stake referred to plant cells, plants and plant tissue in general and not to a specifically defined narrow group of plants, they were not directed to plant varieties as defined in the new UPOV Act.[43] As to the interpretation of the claims the Opposition Division pointed out that they must be interpreted in the light of the description and if the description clearly teaches a broad applicability of the invention to a large group of plants it is not permissible to interprete such a claim narrowly as referring to a plant variety. On the other hand, it was stressed, if the description taught a narrow applicability of the invention to a specific plant variety then a corresponding claim of a generic form would be inadmissible as being directed to a plant variety.[44] This practice of the European Patent Office is highly needed in order to secure the extremely important generic protection of new technologies for producing plants, which cannot be offered by special protection of new varieties of plants provided for along the lines of UPOV Convention. The latter always relates to a specific plant variety only.[45]

22. With respect to the patenting of plants mention should also be made of a recent decision of the German Federal Supreme Court of March 30, 1993 in a case in which claims were directed to a variety of a "tetraploid camomile".[46] Irrespective of the unique legal situation, as a result of which, in this case, a "plant variety" was eligible for patent protection,[47] our Supreme Court explicitly confirmed that as regards patentability of biological inventions in general and those of plants and plant varieties in particular, the principles which have to be applied, are those which have been developed in respect to chemical substances. Thus, product claims to plants are allowed if the subject matter claimed is unequivocally identified by characteristics, which are apt to external or internal perception. If this is impossible or totally impractical, then the plant can be described by unequivocally distinguishable and reliably identifiable parameters of its characteristics. Should this also prove impossible, the plant can be characterized by its production process, and thus can be granted the so-called "product by process" pro-

---

42 Patent Application No. 84302533.9-Lubrizol Genetics Inc.

43 No. 9.2.4 of the Reasons for the Decision.

44 ibidem, No. 9.2.4

45 As has been correctly pointed out by Crespi, Patents and Plant Variety Rights: Is there an Interface Problem?, 23 IIC 168 ss. (182) (1992).

46 1993 GRUR 651 ss.

47 Until the last amendments of the German Plant Variety Protection Act and Patent Act of March 27, 1992, patents were obtainable for those plant varieties, which were not included in the list of protected taxa annexed to the Plant Variety Protection Act. This was the case for the camomile, when in 1983 the respective patent application was filed.

duct protection, provided that method can be reproduced by the person skilled in the art on the basis of the written description.[48]

23. When the EPO for the first time had to determine the issue of patentability of an invention related to a method for producing a transgenic non-human mammalian animal having an increased probability of developing neoplasms and to transgenic animals produced by said method – the Harvard "Onco-mouse" – not only the issue had to be resolved whether such inventions were generally excluded from patent protection under the language of Article 53 (b) EPC,[49] but also whether the granting of patents for such inventions would contravene the ethical clause of Article 53 (a) EPC. As regards, the exclusions under the specific rule of Article 53 (b), in the 1990 Technical Board of Appeal "Onco-mouse" decision, the principles, which had been developed for inventions in plants were declared valid also for inventions in animals.[50] Thus, the exception to patentability under Article 53 (b) applies to certain categories of animals only but not to animals as such. For the time being, however, it is still an open question, how those categories of animals, namely "animal varieties", "races animales" and "Tierarten" respectively are to be defined. Considering the revolutionary changes which had taken place since the adoption of the Strasbourg Convention and the EPC, the Board even questioned whether Article 53 (b) can be applied at all in respect of animals which are genetically manipulated.[51] The Board of Appeal, however, made it completely clear that the process employed, namely that of rDNA and subsequent microinjection was not an essentially biological one and that product by process claims directed to animals were admissible. Moreover, it is explicitly stated that patents were grantable for animals produced by a microbiological process.[52]

---

48 No. IV 4 b of the reasons for the decision (1993 GRUR 655). The court, however, did not decide on whether the requirement of enabling disclosure in the field of plants could be satisfied by a deposit of propagating material. From its obiter dicta one could cautiously deduce, that that would have been the case had the Court to decide upon (No. IV 4 c in fine).

49 Note: For the interpretation of the EPC, the English, French and German language wording is equally decisive. In respect to the exclusion of "animal varieties" the terms used in those three languages differ, however, as to the substance. Whereas in the English language mention is made of "plant or animal varieties", the French version uses the expression "obtentiones végétales ou races animales" and in German it reads: "Pflanzensorten oder Tierarten". In view of the ever since the Strasbourg Convention parallel treatment of plants and animals and taking into account the English wording, the term "variety" in respect to animals should be interpreted in the same way as in respect to plants, i.e. as a grouping of animals which are largely the same in their characteristics, i.e. "homogeneity", and remain the same within specific tolerances after every propagation (i.e. "stability"). In other words, as a breeding line. Cf. Straus, 1990 GRUR Int. 922; Moufang, Patentability of Genetic Inventions in Animals, 20 IIC 823 ss. (834 ss.).

50 1990 O. J. EPO 476 ss. = 22 IIC 74 ss. (1991) – "Onco-Mouse II".

51 No. 4.7 and 4.8 of the Reasons (22 IIC 82 [1991]).

52 Nos. 4.9 and 4.10 of the Reasons (22 IIC 82 ss. [1991]).

24. The considerations of the Board of Appeal under the "ethical clause" of Article 53 (a) EPC eventually resulted in an instruction that a test has to be performed on a case by case basis, whereby a careful weighing up of the suffering of animals and possible risks to the environment on the one hand, and the inventions' usefulness to mankind on the other, has to take place.[53] In the meantime the Examining Division has performed the necessary tests and found that the benefit to the medical research of laboratory mice with accelerated tumor development prevail over the suffering inflicted, and potential risks for the environment virtually do not exist. Consequently, the patent was granted[54] but is now opposed by a great number of opponents.[55]

25. Notwithstanding the thorough action of the Board of Appeal and the balanced outcome of its considerations, which took into account the concerns of the public at large, as well as the interests of inventors, who are left with absolutely nothing if their applications are refused,[56] this decision raises questions of fundamental importance. While the European Patent Office has a duty here to search for a common morality codex of the Contracting States and to apply their "ordre public", I do feel that it cannot be its task to "invent" such a common codex.[57] For it is primarily the responsibility of national courts to decide what is contrary to morality and public order in their countries. The situation will be a different one,

---

53 No. 5 of the Reasons (22 IIC 83 ss. [1991]).

54 Decision of April 3, 1992, 24 IIC 103 ss. (1993) – "Onco-Mouse/Harvard II".

55 Altogether there are 16 opposing parties, representing some 25 organisations, among others the German state of Hessen and Members of the Austrian Parliament. The decision has been predominantly opposed on the basis of Article 53 (a) EPC, by claiming that the risks/benefits test has not been diligently performed, the potential benefits of the invention being overestimated on the one hand and potential risks for the environment underestimated on the other. Cf. Jaenichen and Schrell, Die "Harvard-Krebsmaus" im Einspruchsverfahren vor dem Europäischen Patentamt, 1993 GRUR Int. 451 ss.

56 Whereas decisions of the Board of Appeal of the EPO are final for the applicant, competitors and other interested third parties still have the option of applying to national courts to have a European patent revoked, provided the grounds for revocation under Article 138 EPC are fulfilled.

57 The position of the Board of Appeal is not yet clear so far. Tendency, however, could well be a search for a common "public order" of the Contracting States. Moreover, even an effort aimed at establishing a common morality standard appears imaginable, notwithstanding all the differences still existing in the contracting states in this latter respect. Cf. Kinkeldey, Die Patentierung von Tieren, 1993 GRUR Int. 394 ss. (398 ss.). Also some commentators explicitly advocate the necessity of this approach, cf., e.g. Moufang, in: Europäisches Patentübereinkommen, op. cit., footnote 22, Art. 53 Marginal Notes 27, 28; same, Patentierung menschlicher Gene, Zellen und Körperteile? Zur ethischen Dimension des Patentrechts, 1993 GRUR Int. 439 ss. (445 ss.).

once the Community Patent Convention has entered into force.[58] In any case an effect of rejecting of a patent application based on Article 53 (a) EPC which would either make the impression that the law maker is freed to adopt laws in the pertinent branch of public law, or practically preempt technological developments in areas in which the legislator did not yet make up his mind, appears economically and legally unacceptable. Such tasks are not within the competence of our patent offices, although they were assigned to some patent offices in former socialist countries. They would certainly contradict all market economy principles!

26. I therefore very much concur with the reasoning of a recent decision of the EPO Opposition Division of 25 February 1993[59] in which a patent for a herbicide-resistent plant, invented by Plant Genetic Systems from Gent in Belgium, was at stake. Deciding on a opposition of Greenpeace, the Division refused to apply the "weighing up" test analogous to that designed by the Board of Appeal in the "Onco-mouse" case. With reference to the Examination Guidelines it stated that that decision did not supercede the general approach that Article 53 (a) EPC is to be invoked only in rare and extreme cases. The "Onco-mouse" decision was such an exceptional case relating only to the patenting of animals, moreover – in the specific case in question – of animals used as test models in an assay where suffering is unavoidable.[60] According to the reasons of the decision, opposition proceedings before the EPO are not a proper forum for discussing the pros and cons of the genetic engineering of plants in general or the plants in the particular case. In view of the fact that Greenpeace claimed the necessity to perform a risks/benefits assessment of the invention in question, the Division also recalled that the assessment of risks and the consequent regulation of the exploitation of the invention were a matter for other bodies to consider and were outside the scope of the EPO. Moreover, the Opposition Division pointed out that even if potential risks had to be taken into account in this context, the burden of proof in the opposition proceedings was with the opponent. But Greenpeace has not been able to prove, or at least render highly plausible, his risk's allegations.[61] It appears to me that this concept represents a correlative to the "Onco-mouse" weighing up test. This correlative is necessary to retain Article 53 (a) its exceptional character within the context of the EPC and also within our legal order and leave the EPO the manoeuvring space needed in order to adequately cope with borderline cases. However, without being continuously harassed by erratic desires

58 As long as an "autonomous European ordre public" and "morality" do not exist, the EPO should, in this writer's view, refuse an application on the Grounds of Article 53 (a) EPC only if the publication or exploitation of the invention at stake would be contrary to "public order" or "morality" of all the Contracting States. Since the applicant, who would uphold his/her application also for those states in which such a violation would occur, would take the risk of national revocation of his/her European patent. Cf. for details Straus, 1990 GRUR Int. 919.

59 Application No. 87 141.5, 24 IIC 624 (1993) – "Patent for Plant Life Forms/ Greenpeace."

60 No. 3.7 of the Reasons for the Decision (24 IIC 621 [1993]).

61 No. 3.12 of the Reasons for the Decision (24 IIC 623 [1993]).

of various social groups. If justified, the desires of the latter could and should be brought to the attention of the legislator and should be adressed and regulated outside the patent law field. In the present discussion only a few commentators[62] have realised that what has been designed in the "Onco-mouse" decision as the weighing up test goes far beyond the issue of patenting animals or plants. If the patent offices had to examine applications, e.g. on the benefits/risks basis and had to take into account economic implications in this context, patents barely could ever be granted for any military equipment or parts thereof, most chemical compounds, many pharmaceuticals etc.[63] If that should be the future tasks of our patent offices, this should be clearly settled in law.

V. Concluding Remarks

27. The crucial understanding that patent law cannot be entrusted a general police function aimed at determining what should be allowed and what prohibited in areas of new socially sensitive technologies without doing harm to the entire technical and economic progress seems to be shared by the EC Commission, too. This is implicitly demonstrated by the fact that the Commission, which in October 1988 proposed a Council Directive on the Legal Protection of Biotechnological inventions,[64] was not willing to accept most of the amendments passed by the European Parliament with regard to this proposal in October 1992.[65] Whereas the intention of the European Parliament was to assign patent law functions which until now, and with good reason, were strictly the preserve of other parts of the legal system, the Commission decided only to bring into the proposal some clarifications, which can be regarded as being in line with the views traditionally expressed in the legal doctrine and with basic principles governing patent protection since its very beginnings.[66]

28. As regards the so-called "ethical clause" of the patent law, the amended proposal specifies that the human body and parts of it, processes for modification of genetic identity of the human body for a non-therapeutic purpose which is con-

---

62 Cf., e.g. Byrne, Patents for Biological Inventions in the European Community, 1993 World Patent Information 77 ss.

63 Byrne's comments, ibidem, are exactly to the point: "Why stop at biological inventions? Article 53 (a) applies generally. Guns, bullets, bombs and gases such as tear-gas, are specifically designed to inflict harm on humanity or animals. The grant of a patent for a new silencer for a hand gun might be no less objectionable to concerned citizens than a patent for Harvard Mouse or Hairy Mouse. Article 53 (a) seems like a device timed to explode randomly in different directions with consequences unforeseen when the EPC was agreed to by the contracting parties."

64 Doc. COM (88) 496 final – SYN 159, O. J. No. C 10, January 13, 1989, p. 3.

65 Doc. PE 161.995.

66 Doc. COM (92) 589 final – SYN 159 of December 16, 1992.

trary to the dignity of men,[67] as well as methods for modifying the genetic identity of animals which are likely to inflict suffering or physical handicaps upon them without any benefit to man or animal, are not eligible for patent protection (Article 2 [3]). Having regard to the debate in the European Parliament, the Commission in the Explanatory Memorandum has clarified that under "parts of the human body" are to be understood only those parts which are actually in the human body. Thus, the amended proposal is not aimed at excluding from patent protection any subject matter for which patents have already been granted at the European Patent Office.[68] In respect to the patenting of animals and plants, the amended proposal has maintained the exclusion of plant and animal varieties from patent protection, but has explicitly conceded that plants and animals as such are and remain subject matter eligible for patent protection (Article 3). This holds true also for plants and animals and even plant and animal varieties as direct products of a patented process (Article 10 [2]). More deplorable than the fact, that the exclusionary provisions of Article 53 (b) are upheld in the amended proposal, is the intended introduction of a "farmer's privilege" into patent laws of the EC Member States (Article 13). The latter constitutes a clear breach of all traditional principles governing the patent law and seem to me untenable both for legal as well as for economic reasons.[69]

---

67 Thus, as stated in the Explanatory Memorandum, this exclusion from patentability does not cover the processes used in somatic gene therapy (loc. cit., footnote 66, p. 8). Even germ line therapy, which until now is neither subject matter of patent applications filed nor approved by any authority worldwide, if applied for therapeutic purposes, which for the time being is definitely forbidden for instance under the German Embryo Protection Law (see for details Straus, 1992, GRUR 255 ss.), seems to be allowed under the aspect of ethics. As demonstrated by the proposed Article 8, however, Article 2 (3) (b) of the proposed directive by no means makes these inventions eligible for patent protection, if the requirements of Article 52 (4) EPC are not met.

68 The Explanatory Memorandum reads, in this respect as follows: "With regard to the unpatentability of the human body or parts of the human body, the Commission wishes to make it quite clear, in keeping with the discussions that have taken place within Parliament, that 'parts of the human body' per se means parts of the human body as found inside the human body. It is important that this be spelled out so as to remove all possible ambiguity with respect to the position of certain products or parts of the human body which are already covered by patents granted in connection with the development of medicinal products . . ." Then several European patents are indicated, e.g. for a human lymphoblastoid cell line (EP No. 0113.769 B1), and for a gene sequence coding for human relaxin (EP No. 0096.839 B1). Loc. cit., footnote 66, p. 7.

69 It has to be noted that the expenditures for, e.g. the licence share in certified seeds in Germany are marginal as compared to the expenditures for seeds, herbicides or pesticides and especially to subsidies. Cf. Priebe, Agrarpolitische Rahmenbedingungen einer umweltverträglichen Landwirtschaft, in: Albrecht (ed.), Die Zukunft der Nutzpflanzen – Biotechnologie in Landwirtschaft und Pflanzenzüchtung, Frankfurt and New York 1990, pp. 53 ss.; Straus, 1992 GRUR 265. The negative effects of a "farmer's privilege" are foreseen by agricultural economists also for animal breeding. Cf. Henze and Zeddies, Kosten-Nutzen-Untersuchung im gewerblichen Rechtsschutz in der Tierproduktion, Hohenheim 1993, pp. 140 ss.

29. Patenting in the field of genetic engineering has certainly been and will remain for some time to come a challenge for the EPO and other patent offices in Europe. Until now, according to my understanding, the European Office has mastered that challenge in quite a remarkable way. It is up to the patent law community to make the general public and politicians the functioning of the patent system, its aims and legal consequences understandable. But it is also the duty of politicians and the public at large to make all efforts in order to understand those mechanisms and their true legal consequences. So far remarkable defficiencies exist and should be overcome. The earlier, the better!

30. I would like to conclude with a remark which Professor Sidney Brenner, the world-famous British researcher, speaking on the problems associated with patenting genetic engineering breakthroughs, recently made:

> "God never patented evolution, but he kept it
> a good trade secret."[70]

Our Scientists have been able to decode a good deal of it and revealed its secrets, and the US and European patent systems have both played a major supportive role. The designations of inventors on genetic engineering patents read like a role-call of the Nobel prize winners over the last 15 years. We would not have had anything approaching such openness had it not been for the patent system, of that there can be no doubt.

---

[70] Loc. cit., footnote 10, p. 146.

Remarks on "Technology Transfer and the Genome Program"

R. G. Adler

Thank you for this opportunity to present perspectives from the National Institutes of Health. The conference organizers kindly have permitted me to submit a paper even though I cannot attend the conference in person. The recent birth of my new daughter precluded my traveling to Germany. This paper will summarize the status of patent applications filed by the NIH, one of the world's largest and most celebrated biomedical research institutions, that encompass roughly 6,500 novel gene sequences, of which about 60 % do not suggest a putative biological function. My remarks also will cover several generic challenges that contemporary biomedical research poses for NIH, publicly-supported research agencies around the world, their academic grantee institutions and for the biotechnology industry.

With the help of computers and robotics, research has expanded – exploded, really – into several areas of scientific inquiry that produce massive volumes of sequence or structural information well before associated functional data is available. That structural information has unquestioned importance to the research community. Also, some of the identified molecular structures ultimately will have great commercial value and relevant patent protection at some point will be very important in encouraging expeditious product development. The technology transfer issues arising from the discovery of potentially powerful new molecules through rapid sequencing technologies – discoveries not yet tied to a biological function – were inevitable from the moment we embarked on a large-scale, aggressive effort to sequence the human and other genomes.

As was the case for other areas of research and patenting in the past two decades, involving monoclonal antibodies, recombinant DNA, plants and transgenic animals as a few examples, collectively we are trying to gaze into the future without a crystal ball to tell us how gene sequence technology will be developed commercially and how NIH and its grantees can best facilitate the transfer of this technology. We also must ensure that fundamental knowledge is available to the entire scientific community. Deliberations by many institutions and organizations in the U.S., here in Europe and elsewhere will provide a framework that will help society deal with the commercial applications of emerging technologies. Recently, for example, the U.S. National Academy of Sciences sponsored a conference on the commercialization of transgenic animals. Industry still is stepping forward haltingly more than decade after the creation of transgenic animals in the laboratory, and important questions have been raised by the research community about access to related scientific discoveries. This debate will provide useful guidance for commercial development and science in the field of genomic research which, in a sense, is relatively newer on the research scene, just five years into the Human Genome Project.

The early public debate about when to file patent applications on gene sequences and what to do with issued patents is virtually unprecedented and very important. We all are stakeholders. As Dr. Bernadine Healy, The Director of NIH, noted in her September 22 testimony before a Subcommittee of the U.S. Senate Judiciary Committee:

> Despite the controversy engendered by the NIH's patent filing, through the legal and policy debate flowing from those filings, we will inevitably move closer to understanding how best to translate these rapidly growing numbers of gene-related discoveries into benefits for the public. No outcome is forced by the NIH and (Department of Health and Human Service's) interim decision to file for patents, and the patent office is in a position to engage an issue of broad relevance.

I refer you to that testimony and to Dr. Healy's August 27, 1992, article in the New England Journal of Medicine, for a very cogent discussion of many complex and inter-related policy issues associated with technology transfer from genome research. Because this conference relates to patent law issues, I want to refer to a more technical paper of mine from Science.[1]

NIH publishes the gene sequences for which it seeks patent protection and makes actual DNA clones available. Another important point to consider when assessing the impacts of patenting and licensing is that, under NIH guidelines, NIH inventions are available freely for academic research without any concerns for infringement. Perhaps we should treat separately the uses of patented inventions for research purposes on the one hand and for diagnostic or therapeutic applications on the other hand. The dissemination of scientific information also may be a separable concern. Preserving appropriate access by the academic community and industry to research tools, whether one is speaking of transgenic animals, cell-surface receptors, genes or other biological materials, and regardless of who discovers them, is a critical challenge. In this broader context, the treatment of gene sequences is but one example or case study. The treatment of research tools clearly will benefit from continuing discussions by Government agencies, the scientific community at large and industry.

As one approach to facilitating access to research tools, my office at NIH and leaders of the academic technology transfer community in the United States have been developing a proposal for the establishment of a Uniform Biological Material Transfer Agreement, or "UBMTA" to facilitate the sharing of research tools. This document contains common denominator terms that will reduce the need for negotiations when materials are transferred between Government and academic institutions. It will discourage the over-reaching attempted by a few institutions,

---

[1] Adler, R. G.: Genome Research: Fulfilling the Public's Expectations for Knowledge and Commercialization. Science 257: pp 908–914

that is, "if you discover anything of commercial value with my receptor that new invention is mine, too." The UBMTA could be implemented in a treaty fashion that will minimize the need for signatures by scientists, counter-signatures and the unnecessary involvement of lawyers in this aspect of research. Material transfers involving industry on the one hand and government or academic institutions on the other hand, however, have some different considerations. Perhaps this approach might be helpful in Europe as well. Also, in the U.S., unlike in Europe, we must monitor how the research exemption from infringement is treated by the courts, with an eye toward seeking remedial legislation should that become necessary or appropriate.

Turning now to the status of the NIH gene sequence patent applications, the first-filed patent application of June 20, 1991, was converted into an application for a Statutory Invention Registration, rather than for a patent, that claims only the rapid sequencing process that NIH hopes will remain in the public domain. The U.S. Patent and Trademark Office issued an Examiner's Action last August in the second patent application, itself filed February 13, 1992, that also claims the gene sequences disclosed in the first-filed case. NIH responded to the Patent Office in February. The third patent application was filed on September 25, 1992 and the Patent Office has not acted on it.

NIH filed its response to the Examiner's Action to answer patent law questions about the sequences generated in connection with a large-scale, international scientific program. For example, an important question is whether the claimed sequence subject matter is patentable, without certain knowledge of the biological function of the products of such sequences. A related, but more important, question is whether and to what extent gene sequence claims will remain patentable to others in the future who discover relevant biological functions after the sequences (or partial sequences) have been published. The answers may vary depending upon how much of a gene has been sequenced or upon how much of a gene's biological function is apparent from its structure. Additionally, a recent court decision by the U.S. Court of Appeals for the Federal Circuit in *Fiers v. Sugano* must be considered carefully in the context of when gene-related inventions actually are made in the patent law sense. This case also has implications for protein research and other areas of bio-medicine.

In responding to the Patent Office, NIH also sought to keeo options open for the licensing of patent rights to gene sequences when appropriate – if such sequences are patentable. The Bayh-Dole Act of 1980, P.L. 96-517, states that "[i]t is the policy and objective of the Congress to use the patent system to promote the utilization of inventions arising from federally supported research or development." Note the Bayh-Dole was enacted before we had to consider the reality of transgenic animals, gene therapy and the Human Genome Project. Some have suggested that this statute and its implementing regulations need to be revisited. I understand that the Max Planck Institute for industrial property and competition law in Munich will convene a separate conference next January about industry-academia relationships and the transfer of technology. Perhaps our basic purposes in conducting research and transferring technology would benefit from clarification or reaffirmation through such meetings.

In implementing Bayh-Dole and the Federal Technology Transfer Act, NIH has developed a socially responsible patent and licensing program to facilitate the development of health care products as part of our mission to extend healthy life and reduce the burdens of illness and disability. We license inventions exclusively or nonexclusively, pursuant to statute and regulations, as appropriate. Because of concerns expressed by some companies, I reiterate a statement by Dr. Healy to the Judiciary Subcommittee, that NIH is amenable to not enforcing patent rights that issue to genes of unknown function, effectively dedicating them to the public, if the public interest warrants that course of action.

We don't yet know, however, which DNA sequences ultimately will become commercial products in their own right, perhaps in diagnostic, gene therapy or antisense applications. Some will only be research intermediates. It also is unclear as to what kinds of patent rights in gene sequences actually will be of importance to industry. In this regard, therapeutic method of use patents simply are unavailable in some of the world's major health care product markets. Moreover, *Science* magazine reported in January that a California biotechnology company, Incyte, has filed a patent application similar to those of NIH. What other companies or institutions around the world have filed similar cases. The extent of commercial interest in gene sequencing research is unquestioned; I understand that Dr. Craig Venter, of the new Institute for Genomic Research, also recently formed a commercial linkage with a major pharmaceutical company. It seems to me that corporate sentiments reported by the media are more critical of the pending NIH patent applications than what some of these companies apparently are saying behind-the-scenes.

To the extent that NIH and its grantees continue to be involved in early-stage gene sequencing, and the private sector is taking a very active and expanding role in this field, I do not see how we can afford a double standard for patenting, one for the public sector and one for the private sector. Certainly the Bayh-Dole Act and the Federal Technology Transfer Act of 1986 to not appear to contemplate one. These laws were intended to ensure a return on the public investment in fundamental research through the development of useful products resulting from Federally-funded research.

NIH is concerned that we learn from our analysis of technology transfer issues as applied to gene sequencing in order to anticipate areas of research that may raise fluture uncertainties about patenting and licensing policies. For example, discussion might appropriately focus on additional areas of research that generate large numbers of molecular structures of academic interest before their functions have been elucidated. Protein research and structure-based (or rational) drug design raises similar patent and licensing issues.

We also might well reflect generally on the question of how to facilitate commercial development in other fields of biology and medicine that are being created through Federal support of biomedical research. For example, how will gene therapy be commercialized and delivered to patients what should NIH and its grantees and their counterparts internationally do to facilitate the translation of this research into corporate commitments for product development?

What is clear today at least is that a resolution of patent, licensing and other technology transfer issues is of great importance important to Government, academia and industry in several countries. Given the rapid pace of technological development for gene sequencing and for other exciting and promising areas of research, we must carefully consider where biomedical research is heading to ensure that our policies and this technology work together.

Discussion: Session 1

Sahai: I would like to address my comments to Herr Straus. Herr Straus went to great lengths to explain the rationale behind the patent system and then to explain about the peculiarities that are being considered in Europe vis à vis the patent system. There are a few things I'd like to point out in this direction. The fact that patents come out of the concept of industrial property in industrial countries, has been validated by the history of the patent system. What is disconcerting at present, is the attempt to harmonize and globalize patent laws through international conventions like the GATT negotiations, the Biodiversity Convention, and I'm sure subsequent conventions. It is also well-known that the patent system which each country has adopted has been commensurate with the state of its technological development. It is not surprising, therefore, that a G7 nation like Italy acknowledges product patents in the chemical and pharmaceutical sector as late as 1982, and Spain as late as 1987. This is only to illustrate that the technological state or what the country feels is its technological state, determines to a large extent the degree of patent protection that it is willing to offer. In this light the present effort to harmonize patent laws with developed countries through international negotiations, is not compatible with the spirit of patent systems that Herr Straus elucidated. There's another aspect that I would like to point out: At these international negotiations, some countries are pressing for general acceptance of several aspects of patent laws, for example, compulsory licensing. Canada has had to shed its compulsory law scheme almost as a provision for joining NAFTA. Such clauses would severely restrict the dissemination of technology in developing countries which Herr Straus announced as one of the functions of the patent system.

Straus: I was assigned a very specific task to talk about the situation in Europe, and as I have seen from the program, you will be speaking about the issues you have addressed just now. But of course I shall say a few words about your remarks. It is certainly true that there is a basic principle of a patent law and all intellectual property rights. They are limited to the territories where they have been granted, and therefore, countries not having such standards of protection or even no protection at all are completely free to use the patented technology, if they wish to do so. That applies to all countries. That means if something is patented in Europe it can be used in India or in any other country. That applies even to seeds and so forth. I was not able to address all issues, but this is a very simple one. If you buy seeds from a company which is located in Europe and you transfer that to India, nobody will try to interfere with that. Another issue is of course the question whether the attempts to impose high standards of protection in developing countries via GATT is something that should be supported or not. But when discussing this issue, one has to take into account the changes which have appeared during the last decades. We have now an internationally globalized production, globalized research, globalized trade. And within this international community, it is virtually impossible for states which claim to be a part of that globalization process to take the position that they may be completely free in what to offer in the field of intellectual property. And allow me, Frau Sahai, maybe one last remark. I have read with quite a modest satisfaction, a report about your country just recently in the Monitor, which is a production of a group in Holland, indicating that Indian companies engaged in plant tissue production have problems because there is no protection for them. And what they do is they try to keep everything secret. I am not a fetishist of patents, but I am claiming to be somebody who is trying to estimate the situation in as objective a way as possible.

Hofschneider: Well, we are starting out already with very global problems here at the beginning. May I suggest that we assure that this discussion will come up during the forthcoming meeting and certainly at the round table discussion. So, I would suggest that we focus a little bit on the contents delivered in these presentations this morning.

Garthoff: Let me comment on one of the issues that was mentioned in the paper by Adler. He comments that private companies are engaged in some kind of cooperation with gene sequencing companies and he's saying, o.k., that's some kind of double morale. But let me point out that the first step of course was taken by NIH, starting off by patenting DNA sequence fragments. And if you do this with functionally unknown sequences, there's only one way for private industry to do the same and try to tie up with them. For private industry in the past it was always the issue that you have to have some function behind it. And the gene sequences being patented, coagulation factors, for example, have some function and therefore, it is of interest for the industry. If other companies turn up and cooperate with gene sequencing companies, this is not a kind of double morale, it's just an issue of worldwide globalization. You just have to protect yourself; you have to make sure that in all countries you have a chance to market and trade.

Vogel: Well, I have difficulties commenting on this, because I didn't write Mr. Adler's paper. But it is a just remark on this paper.

Beier: Mr. Adler was a fellow at the Max-Planck-Institut 10 years ago. If we had known what he has done now, then we would have discussed the problem with him. He's a very good lawyer. But, to be very short – I think that these applications are filed in a too early stage of the research and development process and should never be patented. Mr. Straus remarked already that we have more difficulties in Europe with our very broadly interpreted conditions of industrial applicability. But, in my view, it is not a ready invention – "eine patentreife Erfindung". And I think this will hinder patenting in Germany and in Europe. If Mr.

Straus is of a different opinion, this demonstrates that we at the Max-Planck-Institut have sometimes different opinions.

Straus: Well, I would not say that I have a different opinion, but I would just emphasize one thing which we should not forget when addressing this issue. I mean it is a simple fact that the United States of America invest a few hundred million dollars in that project per year and they don't need any assistance in this regard, but, nevertheless, this should be borne in mind. It's a legitimate action to try to see what will be the outcome. The discussion is there and they may yield too much negative publicity. But the European countries – I'm not addressing others – also invest in this kind of research – just a percentage – and that has to be seen.

Vogel: I also have a question for Dr. Straus. In the traditional European and also the German patenting laws, a clear distinction is being made between discoveries and inventions. Inventions are patentable, discoveries are not. First, I have the feeling that this is not the case in the American system. I haven't found a clear distinction in American papers. In your presentation, you gave a definition of the DNA sequences, maybe cDNA sequences. You said they are not discoveries, they are inventions, because they are made artificially from messenger RNA sequences. Now, a cDNA sequence in my opinion is only the complementary copy of a mRNA sequence and mRNA is available in all functional cells and therefore, I would be very reluctant to call a cDNA an invention. In my opinion it is a discovery. Even an animal, for example, a transgenic animal that contains, let's say, human or other genes is not really an invention, it is a kind of construction, because the important part in it, for example, the gene sequence, is something naturally occuring. Therefore, the entire thing is a discovery. If you could design a completely new animal just from de novo synthesis of DNA and of cells, then I would accept this as an invention. Could you say something about the difference between inventions and discoveries, using these examples?

Straus: I have heard the remark this morning that I have delivered a kind of a lecture, but I see, much more lectures arc needed, of course, in order to explain all issues linked to the patent law. I may address first, Dr. Vogel, what you have said about the U.S. situation. Of course, the U.S. law at the same time refers to inventions and discoveries. This is not meant as a discovery in your sense, but means something which is not a product or a law of nature. Now, coming a little bit closer, it is very important to understand the patent law definition. What is really found in nature? What is a product of nature? And if you say, let's start with the transgenic animal, you, for the time being, cannot find or discover a transgenic animal in nature. You have to solve that problem with an invention which is teaching how to do that. And that was the aim of giving these kind of provisional explanations this morning. Now, with the transgenic animal there should be no problem to say that this is not a discovery. The problem with the genes and the other substances – I would call genes substances – found in nature is an old one. And lawyers have to solve problems not invented by them, but by practical life. For example, a scientist in industry who had discovered new natural substances approached the patent offices and said, well this is Lactobacillus bavaricus, this was the way how I got it and that was something previously unknown, nobody knew that, and it was in a very complex environment. The genome too is in a complex environment, although becoming more and more mechanically explored and, let's say, ordered. And at that time at least the courts in our country and also in the United States, but with considerable differences, said that if something was in nature but as yet unknown, undiscovered, and if somebody is the first to offer an instruction how to get that and how to reproduce that, this is an invention. In other words it is not only a question to find a certain spore produced by a strain in the sewage of Calgary. You have to find the strain, you have to extract and isolate the pure substance and you have to say how to do it. Professor Beier may be somewhat reluctant to mention that 10 or 12 years ago we both had some thoughts on that very problem, how to offer something to the scientists in order to be a full-scale participant in commercial life. And we said that even if something was first discovered without really knowing its industrial applicability, the scientists should be offered the opportunity to file patents, in order to be offered an opportunity to deliver the information needed to overcome the borderline between a discovery and an invention later on, during the patent granting procedure. The scientist should say what it is, how to reproduce it and for which aim it can be used. And that is the problem, as I understand the application, that for most sequences the only hope is that they could be used as probes. But if he could say "You can use it as a probe, in order to detect a specific gene" there would be for us lawyers in the fields of patents, – without now taking into account other possible aspects, – no problem. But the problem is that they don't know whether and for which genes these sequences could be used as probes. I hope this was at least a part of the answer.

Rowland: In any event, I did talk to Reid Adler about the patenting of anonymous DNA sequences, but I think it's unfortunate that we create fictions in the law about why we file patents. It doesn't encourage invention, not in biotech, certainly. But it does encourage investment. Now, we have seen, over the years in the United States, an enormous amount of investment in research in biotechnology. And it doesn't really result in dissemination of information because in the 17 years in our country and, perhaps, the 20 years here, no one has really been very interested in what the patent has to say. But in biotech, for reasons that I have never quite understood, it became the paradigm to publish. The rationale was that in order to attract good scientists, you had to allow them to publish. The companies that originally started were from academia, were run by academicians and they just carried over the lab from the university into the corporate environment. And so it has become pretty much standard that any new company continues to publish fairly rapidly. The fact that you're able to file patents before that doesn't really affect all that much the publication. Adler was faced with the fact that NIH was about to publish a large number of sequences and I will be talking a little about our patent office in the United States and where they're going as to patentability. And he was concerned that if he published partial sequences he would then preclude anyone from getting a patent on the entire gene. And his hope was that if he could get patent rights, the government could then provide the proprietary position to a company which wanted to develop the protein expressed by the total gene of which he had disclosed the fragment. And that's how that all got started in a very naive way, thinking that he ought to try to protect these sequences. And the fact is that he used a genomic library, from neuronal cells; you could use these fragments to see what other cells also express these proteins. It's useful to follow what cells or what cell types and what level of differentiation also share this property within a neuronal cell library. That's the history of what happened. Part of the explosion was that Jim Watson got very upset with the idea of filing, that they had failed to get permission from the highest authorities at Health and Human Services who got very upset by the bad publicity and came down very hard on Bernadine Healy and Adler. The upshot of it is that we have a fairly simple patent situation and that under the standards the group is using, the sequences are not patentable, if that standard is correct, which I don't think it is. And so we're now in an interesting situation where I think we will develop rules. And I would add that the court again, I think, wrongfully, is pretty much saying in this field everything is patentably that's novel. But all you get is what you did, which is very useless. So in the United States, even if Craig Venter got his patent he would only have those fragments with those particular nucleotides and nothing more.

Straus: Yes, I just make a small comment. I, of course, am not so familiar with the Venter problem. But I definitely disagree with your first comment. I'm sometimes a little bit surprised that people engaged in patenting and the day by day work underestimate the attractiveness of patents to scientists and technicians. It would be wrong to say that without patents there would be no inventions. But I think many things are invented because we do have patents. And not only the investment after the invention is incented by that, but also the invention itself. This is the first remark and the second remark, I definitely do not share your view that there is no attraction coming from published patent applications and specifications. This is not true. Especially in the field of biotechnology, I could tell you from your country that you have a patent on erythropoietin; but you have a number of follow-up patents, e.g. from the University of Washington in Seattle – indicating new methods, using different vectors based on the publication of that first patent. So, I would be a little bit hesitant and cautious to underestimate the value of the patent system also in this particular area.

Berman: Prof. Straus, you have mentioned Article 53a EPC. If I am not mistaken, this is the article that prohibits issuing a patent that may be contrary to Public Order (= ordre public) or that encroaches on principles of morality. If I understood you rightly, your opinion is that the European Patent Office must not be burdened with such issues as: *morality, ethics* and *public order* and it is for the legislator to deal with and decide on these issues. I must say that I personally am of the opinion that the crucial question, whether an invention is contrary to public order, should be challenged and answered already at the very first steps of examining an invention and on the factual merits of each individual case. I do agree that it is for the legislator to prescribe the policy and the normative framework, but there must be a discretionary body, entrusted with the examining of an invention, also for the purpose of answering the vital question on a case to case basis, whether there is a clear and real danger of encroachment on public order. These are factual questions, varying from case to case, and can't be dealt with in advance by the legislator. The Patent Office authorities have discretionary powers and in my opinion, they are the first to understand and foresee the effects and affects of each specific invention examined by them. I am of the opinion that Article 53 (a) EPC provides for the most efficient way to challenge inventions also on the merits of "morality, ethics and public order" and it does not mean that the EPO is to become a sort of police force! Wouldn't you find the regulations in the EPC a proper solution? Doesn't it ease the fears of the scientists? What other effective means would you suggest to provide for protecting public order and safety in such specific circumstances, discussed here?

Straus: Of course, I was not exhaustive in that particular issue, too. What I was trying to say is that one has to bear in mind that the patent is a piece of paper unless you are allowed by law to use the invention which is patented. If you violate the law by the exploitation of such an invention, you may not do that. So this is the first thing one has to take into account. The second is, I didn't say that the patent offices shouldn't be bothered at all. But what is addressed there is something which should first be addressed by the constitutional principles and by other laws – in the public law sphere – and not by the patent office. The patent office is not there to decide what technologies are appropriate, or even to evaluate whether there is an economic rationale behind them. I didn't mention that Greenpeace presented (Dr. Gugerell will be talking about that) data showing that farmers in Sweden were against the introduction of herbicide resistant plants. Now that may be true, but what should the European Patent Office do about that? If the Swedish legislature was of the opinion that herbicide resistant plants should not be brought into the market – in Sweden – then this should by prohibited. There is no problem. How can a patent office try, e.g. to find out whether an engine for a jet plane should have a thrust of 4000 lbs. or 6000 lbs. If you have 10 000 lbs. maybe you can power an airplane for 1500 people. Should the patent office think about, whether it is too risky to have airplanes with 1500 people taking off? I mean, this is something which would really somehow sink the whole system if we tried to assign such tasks to the patent offices. And I must tell you, I was hesitant to be that determined in this particular point. But just a week and a half ago I met a person with a very high status in our society and he doesn't know very much about patents, but he said: "If you tried to tackle these things in the patent office, it would be detrimental because, on the one hand, the legislature would somehow feel free from the duty to regulate these issues despite the necessity of such rules, and on the other hand, you will preempt technological developments without providing the opportunity, because if the European Patent Office will say that herbicide resistant plants are contrary to the morality or the public order, what will be the next reaction?" They will say that this must be a wrong technology.

Gugerell: I would like to expand a little bit on what Prof. Straus has just said. Also, as a response to your remark. I do not believe that considering ethics in the patent office is a very economical way of doing these things because the patent examiner who is, at least in Europe, a technically trained person, does not have the necessary competence for doing it. Of course, you can say we have to recruit people who are capable of doing it, but at least at the moment, it's not the case and if you take the conditions of the onco-mouse case into consideration (and I will come back to that this afternoon), the board told us that we have to consider the risks of the given invention to the environment. In the case of the onco-mouse it was easy to decide on that. But if you take a case like herbicide-resistant plants and the patent examiner should now assess the risk when the plant is put into the field, – there is no way for him to get the necessary information to do that. Another criterion for that decision was the benefits to mankind of the given invention. That's also something which you can assess in general terms, but you never know the particulars of the given case and maybe at the time the examiner will assess that it is beneficial the inventor may already know it is no longer beneficial because there are side effects which are unbearable. So, I would say that the assessment of these things and consequently the assessment of ethics in general is not an appropriate way for the patent office to act.

Waugaman (Duke University, North Carolina, USA): I want to a comment on Dr. Adler's use or invocation of the Byam-Dole Act of 1980 in the U.S. This legislation was passed by Congress to take away the confusion that existed at that time concerning the ownership and the right to develop inventions made at universities where the research was supported by government grants. That legislation made it clear that the individual universities were responsible for exploiting or developing those inventions and had the right to expect, if they diligently pursued those inventions, to have title and ownership of any patents that might exist. I think it would have been difficult for any U.S. university to take the action that NIH took with the gene sequences. Any U.S. university would have looked at the sequences and the way they were "discovered" and said, "We don't see the utility. We can't see how an investment of our funds would pay off." Universities in the U.S. file patents with their own funds – not with government support. It would have been difficult for any other research institution in the U.S., but NIH, to file those patent applications. And therefore, I'm afraid that the action in itself might set up, in fact, a double standard for this level of basic research which is normally and widely conducted in academic institutions or in government laboratories. And that is the concern for us in universities.

Beier: I would make some additional comments to the question of Judge Berman as to the applicability of the 53a condition within patent offices. We cannot maintain the view that we will not apply 53a at all, namely the prohibition of patenting inventions which are against public order and morality only because the patent examiner or the patent office is not the appropriate person to judge on these questions. I am also of the opinion that it is primarily the duty of the general legislation to set the boundaries; for instance, with a gene technology law. But if the general legislator is late or reluctant to give clear information as to what is permissable in research and development, then I think the patent office must act. We cannot say we will wait up to four years after the parliament decides on a new technology law. The patent office must act and the first instance is the examiner, as a person. We have bodies, even large bodies of appeal and they can get the information needed for a decision. I think these provisions are in most patent laws for many decades and it is simply not possible that a patent should be granted for an invention which is clearly against morality. Even the publication in an official journal would offend morality. One could not say that this matter does not concern us, we will stick to our technical problem; we will examine if it is new and involves an inventive step; its applicability and all other things do not concern the patent office. The patent system is, in my view, also a part of the general legal system.

Grunwald: I think that we heard theses and antitheses and I'm happy to see that Mr. Adler wrote something like a synthesis. We had a first round exchanging very general views and we are warming up for the second, third and fourth rounds to discuss matters in more detail. Let me just talk about one point from the receiving end of a research center: The material transfer agreement. It is true, as Mr. Adler wrote, that we receive nearly weekly such material transfer agreements bound to material or mice or something like that, stating: "This is ours. Whatever you do is ours, too." So my point is that the exchange of information and material in scientific research has to be kept free. We are here to collect information and discuss what is to be done to do just this. To respect the inventions, to protect the interests of the inventors and investors, but on the other hand, keep free the flow of information.

Session 2

Intellectual Property in the Field of Life Sciences

A. Saint-Rémy

I. The Union legislative process

a) Institutional structure

In order to better understand the situation in the field Intellectual Property linked to living entities, one must keep in mind the structure of the European Union Institutions which is based upon the contents of three international agreements:

- The Treaty of Rome, signed the 25th of March 1957
- The Single European Act, May 9th 1986
- The treaty on the European Union (Maastricht), November 1st 1993

The main institutions established by these treaties are:

- European Parliament
- European Commission
- Council of Ministers
- European Court of Justice
- Court of Auditors
- Economic & Social Committee
- Committee of the Regions

b) Decision making Process

The decision-making process starts with the suggestion of a decision made by the Commission to the Council of Ministers. The proposal is accompanied by advisory opinions from the Parliament and the Economic and Social Committee.

Before reaching the Council, the draft decision travels from Commission to Parliament and back several times ("Hearings" in the Parliament), the contents being examined by relevant Committees of the Parliament before being discussed in plenary session and the contents of amendments being negociated between the two institutions.

It then goes to the Council for decision, modified according to the amendments accepted by the Commission and accompanied by the ones suggested by the Parliament but not accepted by the Commission.

Finally, after preliminary examination by the Coreper[1] and ad-hoc committees, the Council issues a "Common position" which is[2] subject to the co-decision/co-operation procedure, established by the Maastricht Treaty, whereby the Parliament can veto the Common Position and again suggest changes based upon a special majority rule.

The final decision is then taken by the Council which is constituted of 12 ministers of the Member-States in charge of the relevant sector (e.g.: agriculture, industry, research ...). Which shows that cross sectoral negociation and integration must – contrary to what happens to national government decisions – take place before discussions in the Council, i.e. during internal consultation of the Commission services and/or at Parliament level.

In the end, the Council issues either a "Directive" which has to be enforced through adapted national legislation or a "Regulation" which is of direct application throughout the Union.

The Commission is then in charge of implementing and following-up the Council decision.

c) Council decisions linked to the applications of life sciences

Three directives have already been edicted by the Council and implemented in the Member-States:

- *Directive 90/679/EEC,* regulating the working conditions of people who are active in laboratories or factories dealing with biotechnology. Its essentially aims at the protection of workers against diseases or accidents linked to the manipulation of genetically modified organisms.
- *Directive 90/220/EEC,* on the deliberate release into the environment of gentically modified *organisms.*

The main features of this decision are:

Scope:	All GMOs – Require Council decision for amendments (Annex I)
Requirement:	For notification and permit for all operations
Part B	Research and Development
	Technical Dossier specified in Annex II. The risk assessment. Competent Authorities Circulate Summaries via Commission to all Member-States.

[1] Committee of the Permanent Representatives.

[2] In some cases determined by article 189 of the Maastricht Treaty.

Part C	Placing on the market All products require notification and written consent within 90 days. Objections may be raised by the Member-States. Decisions referred to article 21 procedure.

- *Directive 90/219/EEC*, which covers the contained use of genetically modified *micro-organisms*.
  Its contents can be summarised as follows:

Scope:	All GMMs. : Classified in Group I (GILSP-Type microorganisms) and Group II – All Others
Operations	Type A – Teaching, Research and Development Type B – Any other operations
Group I	GILSP Principles – Notification of 1st use of facility and of Type B operations
Group II	Notification and authorisation of all operations and first use.

Other decisions are pending at various stages of the decision-making precess, one can quote;

- Novel-Food Ingredients and Processes (COM/92/295)
- Marketing Plant Protection Products (COM/92/914)
- Movement of Medicinal Products (COM/92/382)
- Transport of GMO's (incorporated into transport of dangerous goods)
- Marketing Genetically Modified Farm Animals
- Organic Production in Agriculture
- Biocides

Some comments will be offered further in this paper on a future regulation creating a Community Plant Breeders Rights[3] while the main part will address a draft directive on the Legal Protection of Biotechnological Inventions.[4]

II. European Union and other actors

The European Union did not until recently represent an active actor on the seene of intellectual property since no reference to a mandate in this area had been made in the founding treaties.

---

3 COM/90/347 – OJ C244 – 28. 9. 90.

4 COM/92/589-SYN/69-OJ. C44 – 16. 2. 94.

It is now involved in many ways; one can think of, for example:

- The project of establishing a Community Patent
- The drafting of a directive on patenting biotechnological inventions[5]
- The draft regulation on plant breeders rights[6]
- The complementary certificate on drugs
- The participation of the Commission to GATT-Trips negociations containing important features directly linked to intellectual property aspects of international transactions

The main actors in the field are:

- WIPO: World Intellectual Property Organization
- UPOV: World Organization for the Protection of Plant Breeders Rights
- EPO: European Patents Office (and Court of Appeal)
- National Patent Offices
- National UPOV Offices
- International Court of Justice
- National Courts

III. The draft directive on patents in biotechnology

a) Some history

As early in 1985, the Commission announced, in its white book on Internal Market, its intention to take an initiative in the field of intellectual property to protect biotechnological inventions. Already then, and still now, the intentions was not to really legislate anew but rather to harmonize the application of patent rights in the Community Member States with a strategic objective to contribute to the fluidity of the integrated market.

After a process of internal and external consultation, a proposal was adopted by the Commission in October 1988.

Since then and until December 1992, the text was examined by the European Parliament which proposed more than 80 amendments in total.

A modified text, including the best part of the Parliament amendments was adopted by the Commission the 16th December 1992 and sent to Council for decision.

After discussion in the various committees preliminary to the Council and the Working Group on Intellectual Property, it was adopted as a Common position after further additions and modifications.

In the course of the procedure, the arguments against – and proposed amendments to the Directive were neither of technical or juridical nature but mostly based on ethical grounds.

---

5 See above.

6 See above.

Objections to the patentability of living organisms were first mainly directed towards the "Human body and its parts" but when the discussions developped, patentability of plants and animals were also put into question.

Finally, the aim of the opponents became clear: they were against patents in biotechnology; some seemed against intellectual property as a notion, some indeed even against biotechnology in general.

The Commission and, it seems, parts of the industry have understood the need for compromises with the agricultural milieu including plant breeders and farmers; the "ethical opposition", animal welfare representatives and public opinion in general, hence the integration of many amendments of which the substance did not meet the Commission agreement at a techno-juridical level.

b) The current situation

The Council of Ministers have taken a Common Position in the beginning of 1994[7] which is still the object of difficult negociations between the European Parliament and the Council.

The draft directive will be one of the first decisions submitted to the co-decision process created by article 189 of the Maastricht treaty.

According to this disposition, the Commission has gained a new role as an intermediate between the Council and the Parliament. The Commission has a specific mandate to present compromise proposals to the two other institutions. ("habilitation procedure")

Discussions are taking place mostly around the contents of article 10 of the recitals which, in the Common Position, reads: *"Whereas, in the light of the general principle that the ownership of human beings is excluded, the human body or parts of the human body as such, for example a gene, protein or cell in the natural state in the human body, including germ cells and products resulting directly from conception, must be excluded from patentability, but isolated parts of the human body should not be unpatentable merely because of their human origin, it being understood that the parts of the human body from which such isolated parts are derived are excluded from patentability."*

After amendment n° 3 of the Parliament, it would become: *"Whereas, in the light of the general principle that the ownership of human beings is excluded, the human body or parts of the human body, for example a gene, protein or cell in the natural state in the human body, or isolated from it, including germ cells and products resulting directly from conception, must be excluded from patentability; whereas, however, an invention incorporating isolated parts of the human body should not be unpatentable merely because it uses material of human origin, it being understood that the parts of the human body from which such isolated parts are derived are excluded from patentability."*

---

[7] COM (94.C – 101/94) OJ 101/65 – 16 February 1994.

The current stance taken by the Parliament rapporteur (who has now resigned) is to demand that recital n° 10 be suppressed altogether, which the Commission cannot accept because, in its opinion, this would create a gap in the scope of the directive, install a troubled juridical situation and would also be substantively wrong as well as in contradiction with the Munich treaty and the practice of the Eurpean Patent Office (e.g.: European patents "on relaxin" and "interferon".

The Commission has clearly stated that it could only accept further amendments in one (or several) of three instances, i.e. if they would:

- technically enhance the text
- enhance the global legal and juridical contexts
- constitute an "unavoidable political compromise".

Besides, article 10 of the recitals represents a peculiar case in community legislation: it does not refer to any article in the directive but is a disposition in itself.

This means that national laws being bound to only implement the text of the directive itself the courts would subsequently be obliged to base their decisions on the article in the recital.

Hence, suppressing it would introduce incoherence.

c) The substance

1) Ethical aspects

As mentioned earlier, the focus of the discussions in the Parliament has been on the ethical aspects of the text.

These have mainly to do with the patentability – non patentability of the "human body and parts thereof" and also, but less essentially with the "suffering of animals".

The Commission has always taken the position that ethical points should not be dealt with in the context of the intellectual property system of which it is not the purpose. Besides, the structures in charge of the implementation of the system are constituted of technocrats who are in a very uneasy position when asked to answer questions such as: Does the invention potentially harm "public order"? Or, is laboratory animal suffering properly compensated by the "seriousness" of human diseases to be cured?

Still, notwhitstanding this, the Commission has accepted the need to integrate some "ethical" amendments as a way to political compromise and a means of meeting some public opinion concerns, in line with the Declaration of Human Rights, now a formal basis of the Maastricht treaty.

2) Technical points

Some articles on the Directive and/or the recitals have to be mentioned as departing from the existing systems:

- **Compulsory licensing**

  Article 14 renders it mandatory to licence a patent to somebody who is developing a new plant or animal variety based on the patented object; this for the sake of public interest.
  This article has been considered a concession to plant and animal breeders and a way to achieve a fair balance between plant variety – and patent rights.

- **Patenting animals**

  This has been the subject of much controversy.
  The proposed directive disposes that no object should be declared unpatentable on grounds only that it would be constituted of living matter.
  Beyond this, there is nowhere else in the text a prohibition of animals as objects of patentability.
  This was, after hard discussions, accepted by all Member-States but Denmark, bound by a decision of prohibition edicted by its national Parliament.

- **Reversal of the burden of the proof**

  Article 17 is rather revolutionary in this that it install the reversal of the burden of the proof which has created much controversy since it seems to contravene the principle of "innocence until proven guilt": this disposition – covering only product patents – has been made necessary by the relative easiness to steal, for example, modified bacterial culture of which one metabolite would be covered by a patent.
  This culture representing a "turnkey factory" and being self replicable, the only way to deal with infringement was thought to be this reversal of the burden of the proof.

- **Farmer's priviledge**

  The Parliament insisted on the addition of an article creating a farmer's privilege as exists in the plant variety right.[8]
  This would mean that a farmer would have a right to preserve enough (patented) seeds from his own production to cover his own needs for the following season without paying royalties to the patent owner.
  This disposition is likely to come back on the table of discussion with the Parliament in the course of the co-decision procedure.

---

8 Although considerably modified by the 1991 revision of the UPOV convention.

If it stands in the final version, it will constitute a worldwide "première" in the patent system.

Many personalities, organizations and representatives of the industry have expressed the view that this inclusion would represent a fatal mistake, extremely detrimental to the development of a European biotechnology-based industry.

d) Conclusion

The industry and patents authorities are now convinced having agreed upon the ultimate concessions to interest groups.

They claim that the proposal has gone far enough in that direction.

Most compromises which have take place are about ideological or ethical points. The main problem ahead seems to remain that – weakened as it is – the implementation of the directive will still meet strong reactions in the public, mostly triggered by information from pressure groups.

This challenge can only be met through providing the public with objective, rigorous information both on biotechnology as a group of potentially very useful techniques and about the real nature of the patent system; a system which proposes an exchange between society and the inventor. The latter being granted a monopoly for a determined period of time provided he discloses his invention completely which can be considered a fair deal allowing the inventor to get a fair return on his investment while putting the knowledge thus gained at the disposition of society.

One should also stress the fact that a research exemption is to be found within all existing intellectual property systems.

IV. A regulation creating Community Plant Variety Rights

Since plant cells or plant tissue and indeed whole genetically, modified plants can be the object of either a Plant Variety Right or a patent, we will briefly examine the draft regulation (COM/90/347) establishing a Community Plant Breeders Right.

Suffice, – for the purpose of this paper – to say that this proposal is currently examined by the Parliament ant that the points in discusion are, i.a.:

- Parallelism with the new UPOV convention: since the Union has applied to become a member of UPOV, the proposed regulation will probably need to be re-drafted to become adapted to the new dispositions of the Convention.
- Farmer's privilege:
  here again, the new UPOV convention has considerably modified the contents of the "farmer's privilege", for example by demanding that the interest of the owner of the plant breeder's right on the original seeds be "properly taken into

account". The Commission document will require some further re-writing to accomodate there new dispositions.

- Compatibility with the present state of the directive on patents.
  This draft directive has indeed been modified in 1991 since the drafting of the proposed regulation which probably requires some adjustment.
- Location
  Discussions have already started about the location of a possible office to implement Plant Variety Rights at Union level.

V. Patenting human genes

NIH[9] have been claiming several hundreds of patents on human cDNA sequences of an "unknown utility".

This – and an MRC "retorsion" claim of the same kind – had created turmoil in the research milieu where academics felt that such patents would gravely hamper the flow of scientific information between research laboratories.

After a first refusal from the U.S. Patent Office, the NIH whitdrew their claims, thus settling the matter.

To be noted: during the period of turmoil created by this story, it seems that almost everybody mixed up the notion that the knowing the *in-vivo* utility of the sequence within the genome (which did not matter in view of the patent right) with the description of a new utility which should have to be shown in order to get a patent on the sequence.

9 National Institute of Health of the United States.

Discussion: Session 2

Vogel: First, you were obviously right saying that there are many other complex problems involved. And the second statement you made endorsing the statement which I made in the beginning, that regulations which will be published will meet with a very, very strong reaction from various parts of the public and I think we have to consider these reactions very strongly, in order not to see movements which will kill part of scientific research – not directly by patents but indirectly. I think Dr. Straus has already said that attacks aimed at patenting are meant as attacks against biotechnology as a whole and genome research as a whole. And this is something we have to be very careful about.

Bayreuther: I'm from the local gene center. I have two points that I want to discuss and ask you to comment upon. You said that the human body and its parts will be rejected from being patentable and the point I have is, this is one of these positions without any understanding of biology. If you exclude the human body and its parts from being patentable, then you have to exclude all primates because many of the genes in primates are so similar that any patent that will be issued for a primate gene will apply to human gene. So that's a ridiculous decision. The other thing is, you said that any decision will cause very strong public reactions. There are two types of public reactions that we experience in this country. We had for a long time no way of industrial production of gene products. But we let them come in as imported products. The public was against making the product, but they bought the product. Now, who is overriding the decisions. I think the people that bought the products. The others are just people that are not responsible and did something very, very bad to our science and industry.

St. Rémy: I understand what you say. It seems you're accusing the commission of not understanding biology. Well, in return, I would accuse you of not understanding politics. Tocqueville wrote a few centuries ago that: "la politíque c'est l'art de possible" and it has not much to do with biology I'm afraid. So we were really in a position at a certain point in time – I would call it something like between 6 months and one year ago – that we were bound to choose, either accept or drop the proposal and forget about harmonization in the member states, which was strongly against the clearly expressed will of our industry, which conditions our economic future and implement this reserve of patentability on parts of the human body. And, as I said, and I will be even clearer on that, we, in the commission, strongly disagree with this amendment. Just again, if you find a way out, getting the parliament to accept the proposal without implementing that part of their amendments – I'd be extremely thankful for you to tell me how. Next is, when we talk about public reaction, it does not mean that one should draw a legislation that would please everybody at once and it would be applauded by society 100 per cent. What I was merely saying was, we think we have reached a point of compromise between irrational fears and ideological positions and scientific rigorous assessment of risks. We *think* we have reached a political compromise. That's point a. Point b: once this political compromise gets translated into a law, there will be reactions. I do not say we should avoid reactions. What I said is merely that the way in which we will manage, we will meet these reactions. I do not say we can do without reactions, I say, there will be some and I'm not so sure we are prepared to deal with them in the proper way.

Garthoff: I absolutely agree with my counterpart from science who just mentioned before that apparently the commission is reacting to adverse reaction of public pressure and we had that same reaction in Germany that led us into that kind of gene technology law we had in the past. From our point of view which is that of the industry, both the compulsory licensing and the farmers' privilege are strictly contradictory to what normally was a patent law in the past. It is more or less reacting to public pressure and what we expect to result from that is additional pressure of pressure groups into other directions. So, you're opening up the kind of directives to just an issue of pressure groups.

Lange: I would first like to make a statement from the breeders' industry view and secondly, I have a question for you. First, to the farmers' privilege: I would say more farmer saved seed provision because it is not really a "privilege". But, I think also because the new concept, as you have explained, is against the GATT regulations, special Articles n° 27, 30 and 31. And it is, of course, not in line with the concept which was built in UPOV. And if the European Community wants to be a member of UPOV, this must be harmonized, in my opinion. Under the UPOV system, the reason why the farmer saved seed provision was that the scope of protection was widened. And, therefore, the introduction of farmers' privilege by interpretation came now under a new light and, therefore, there had to be a formulation of this provision. It says that it should take care of the interests of the breeder and therefore requires, in my opinion, a proper remuneration because the user is using intellectual property rights. And the third statement of the *farmers' privilege,* or better said, farmer saved seed provision, I would like to make is that this will create a disadvantage for European farmers. It will not be in their interest. They must have access to the best technology in the most advanced varieties. And I think a fair price will be established by the market. And the breeders now will concentrate their energies on other product strategies and on products with better protection, like hybrids and that is, I think, not in the interest of the farmers. We have seen that some companies have already given up wheat breeding for that reason. The question which I would like to ask: Under the new community plant breeders' rights system, will it be possible to give a license agreement on a country basis? I'm not sure whether this is possible because the community system grants a right to the whole community and not to some states and there might be some – even antitrust – questions to give exclusive agreements to special partners in some countries. I think that it is necessary to have such contracts.

St. Rémy: I will try to answer both questions. First, Dr. Lange's, because it's the easier one. Talking about the antitrust laws at the level of community and the difficulty of granting national or regional licenses, it is certainly a problem, as it is a problem with any community-wide directive or system. I'm not a specialist in that area, so I cannot answer you with certainty, although it seems to me that the community plant breeders' rights would not preclude from national ones. So you could be in the community-wide catalogue and the national one. And then the national one would provide you with an opportunity for national licenses, because if you're in a national catalogue it can prevent somebody from using your variety. That's only a very partial answer. I am sorry to say, I do not have the full one. I have interogated who is in charge of this regulation and he doesn't have the answer either because that must be resolved at the level of negotiations with the council. And talking about the so-called *farmers' privilege* or farmer saved seed provision is just like – you don't call an underdeveloped country "underdeveloped" anymore – it's just a question of vocabulary. It's the same story. What it means is, a farmer can save seeds for the next season without having to pay royalties. Yes, you are right, it is against the interest of most actors, if not all, including farmers – this is also our view. It's just that it seems to me that there is a dichotomy between some actors and their representatives like federations, like some groups in the European Parliament a.s.o., seem to be quite distant from the real interests of their constituencies. But that's not for me to judge upon. We've had to bend to some demands made by the groups we are in negotiation with, i.e. the federations and the groups in the Parliament. Now whether these indeed represent the interest of their constituencies – that's another story and it's not for me to decide upon. Next is, as you can see here, and it will also partly answer your question, that we indeed do not intend to go forward with maintaining Article n° 13, i.e. the farmers' privilege as it was in the parliamentary amendment. We intend to erase Article n° 13 completely and have part of the Article n° 11 which deals with exhaustion of rights. We hope that that can be resolved and we are sure it can. Within the commercial agreements with the inventors and the seed companies or with the seed companies and their distributors it will follow that the reward for the invention of a new seed would be taken care of by the starting price. That's one part of the answer, it's not the full one. Next is, I come back to the structure of the European institutions. It has a, what I would call, a built-in weakness: As I told you, the Council of Ministers is the only decision-making body and it is not a European body. It would not put together the ministers of several disciplines at the level of the Community. On the contrary, it puts together the ministers in one sector, from each of the member states. These ministers do not represent the European interests whatsoever. They represent their national interests. Hence, the results are always compromises between national interests, which is not the same story as if it were interests of the community as a whole. That is my view at least. And it is not abnormal, it is not a mistake in the structure. It is due to the lack of will in the member states and their populations to have a real policy at the level of the Community. That will come with time, I guess. But there is no real will for the time being. Hence, to finish answering your question as much as I can, if you want to lobby, which I certainly strongly encourage you to do, it is not at the level of the community institutions anymore that you should to it. That has been done. I must say, the number of lunches that I have shared with people who wanted to change my opinion is incredible. But it's finished. Unfortunately, I have no lunches to look forward to anymore because the decision is now with the national governments. When they meet in Brussels, it will be your national minister of industry, who will be there. So, if you have ideas to get through – which I guess is the case and which I would strongly encourage, you must do that at the level of your national governments and not at the level of the community anymore. It has left our hands in the Commission and gone towards the Council of Ministers, which, as I said, is not a real European body.

Vogel: Thank you so much. I think your last remarks were very important and very informative for all of us because many of us, at least myself, didn't actually know precisely how the European Community functions and therefore, this information was very useful.

Izhar (Volcani Center, Israel): I wonder whether you are ready to make a comment about the following: What is the European position on the legislation vis à vis science? In other words, we are talking about plant breeding mainly on biotechnology. What is the situation of the legislation vis à vis progress done on science, especially progress in biotechnology and molecular biology? Are we, as society, ready with legislation or are the scientists ahead of the community?

St. Rémy: Thank you for the question, but I'm afraid it's a bit of a philosophical one rather than anything. I don't know of any general, global legislation that could occur or has occured in terms of putting limits to science. All I can say is my own view and my Directorate General's which is that we are certainly as strongly as possible against regulating science, research and development. Science must be free, the more so since it doesn't produce anything but knowledge. That is its aim. Further steps constituted by manufacturing products, multiplying, developing, selling – should be regulated with a view to risk, with a view to what is acceptable to society a.s.o. through legislation on a democratic basis. But in as much as science is constituted by the pursuit of truth and knowledge, we do not think it should be regulated at all. There are ethical committees mushrooming everywhere at the level of member states, sectorily, at the level of the Commission – we even had to put one together. There are some in the United States a.s.o. It certainly is a help for scientists and for cryptoscientists like Eurocrats who are defining research programs. It's certainly very interesting to have the views of such committees or of interest groups or of pressure groups or political parties. Why not have that shed a light on the progress of science and help us to decide upon what we're going to do now? I do not think that it would be a good idea for it to become compulsory through elaboration of law and is not the case yet, to the best of my knowledge.

Beier: I have a remark and a question to our speaker. He says if one would like to lobby for or against a proposal of the EEC Commission one has to address the national minister who can influence the decision of the Council of Ministers. But that is sometimes too late or not possible, especially if the proposal is part of a package of proposals coming from different General Directorates and their Commissioners. I don't like that sort of package deal within the Commission, because it hinders the competent services of the Commission to evaluate each proposal on its own merits. The other thing is that, if I have understood you, the EEC Commission, especially the Director General, which is dealing with science and research will not engage in drafting a directive on genetic engineering but will leave that to national legislation. But what if you intend to propose a research program in biotechnology and in some countries, for instance, in Germany, this kind of research is forbidden by a national gene technology law? I think this would be a point which the Commission should try to harmonize.

St. Rémy: First, I must say with due respect that there must have been a new system instituted in the Commission since yesterday because until yesterday I did not know of any package system. Hence, this is great news. I mean, I will try to learn more about it when I am back in Brussels. I must say that on the contrary, any proposal from the Commission first goes to a very long and very careful process of consultation within the Commission services, i.e. also all and each of the services being in contact with a particular constituency, would then represent the interests and the view of that particular constituency. A proposal like the patent one, for example, was issued by DG3 in charge of industry and Mr. Bangemann. But before reaching the level of being presented to anybody outside the Commission, i.e. Parliament, Council or member states, it has been circulated through the Commission's services which represent in a way different constituencies, like DG12 representing the academic world; like DG6 representing agriculture; like DG11 representing people in charge of environment a.s.o. And the points have been discussed between the different Director Generals one by one and not all as a package. That's answer number 1. Answer number 2 is, and I'm sorry to say again, Professor, there is also no system in the Council, it seems to me because I wasn't aware of the package system there either. On the contrary, I am personally participating on behalf of the Commission in a great number of committees preliminary or part of the Council decision and the points are discussed one by one; we'd spend even 3 or 4 months on one individual point and the Council is perfectly free – that's in the treaty – to approve of a whole directive, to disapprove a whole directive or to change a directive. It is a perfectly free body. Hence, coming back to my previous point, if any lobbying has to occur, what I said when I said it ought to happen at the national level, that's because there's no point in lobbying the Commission anymore, because there is a final position of the Commission. Even if you change our opinion, you won't change the text, because it's been proposed to the Council and at that level it's only through national government that you can have any influence whatsoever. And again, I wish we'd have perhaps more private conversation because I'm always joking at the beginning – sorry about that – but I am not aware at all of any package system. I'm aware of the exact reverse image and I've been around for about 15 years now, so I ought to know.

Rowland: The issue of parts of the human being not being patentable, if I understood you correctly, concerns me from a number of standpoints. One is, I would assume that the genomic gene could be considered a part of a human being and of course that has been the subject matter of patents in the past. Also, there's a very large investment today in the United States and I think world-wide on cells, cell therapy, particularly blood cells where they will take fractions and use them for bone marrow transplants. Furthermore, there's the gene therapy, where you'll modify the human cells and return them to the host. And there are also the efforts to isolate T-cells from cancers and use them as a therapy so that this is something which, at least in our country, has been the subject matter of patents and there's no challenging that being continued and if I understood you correctly, do you mean intact parts or do you mean a part which can be derived – bone marrow could be intact, but if I take cells from bone marrow, it'd no longer intact bone marrow, but it could be considered either a part of a human being or not being a part since it doesn't exist naturally as a part of a human being. And so I'd like you to tell me, how is the word "part" defined in this proposal?

St. Rémy: Well, at least we've tried to. I must say, we're not very pleased with the result yet. But the proposal has not been finalized at the level of Council yet, either. But, a), I understand your concern, we share it, as I said, we've just been bound to compromise for political reasons. I don't see why we would have to hide that. These decisions at the level of Council are political obviously in view of Parliament. And so they're not always entirely satisfactory in terms of either the industrial views or scientific rigor. I do think that there is no doubt that a cell is a part of the human body. I cannot see how you can turn that around. But, if you're talking about what the directive would anyway prohibit is patenting of parts of the human being per se. I guess technical questions on the difference between per se and product by process or product by utility, juridical distinctions like that, I guess have better interpreters here than myself, probably a colleague from the patent office.

Gugerell: I was present at the last Council meeting when this point was discussed and basically these meetings are secret, but generally I would say, it's not only you who is concerned for obvious reasons because the European Patent Office grants patents on genes, also genomic genes. The tendency now in the Council is this that the term part of the human body will be further explained in a *recital* at the beginning of the directives. And there it will be defined that parts of the human body, outside the human body, are not included in that stipulation, whatever that means. But it's not yet finalized.

St. Rémy: Well, I'm a bit sort of split on this point which is a very important one, especially to the pharmaceutical industry. I understand that certain cells could be very useful in the next generation of pharaceutical molecules – everybody knows that. Just one short point is, I am really split: I think that we will achieve in the Considerat to the directive, a correct definition meeting your worries, I guess, otherwise we will not accept the drafting. Only, I am not so sure I would trust the Considerat to be applied by the courts afterwards. And that remains my worry.

Vogel: The discussion starts to become really very interesting because the point Dr. Rowland has made is precisely the point that has upset me – the practice in the United States. This has in fact induced me to convene this meeting and I must say, I repeat it, I am really extremely critical about these points. I think the Americans have started something and the Europeans or other people are forced to follow unless we cannot meet the industrial competition.

Sahai: Did I understand you correctly that natural genes under these directives are patentable because natural genes are always associated with human intervention?

St. Rémy: It's a complex story and it's not so complete yet as Dr. Gugerell has pointed out. We are still in the process of drafting a very delicate text on that particular point. So, I can answer you a) about my opinion, b) about what I feel will be the text in the end. None of this is formal because my opinion is not very important and on the other end, the text has not been finalized. It could be surprising. But, what I mean here is the present state of the reflection, and what it says is, this is aimed at reassuring the industry, to be clear; what it says is, don't worry. The text will say that a naturally occuring gene in a human being is not patentable. Don't worry about that because you will have some work to do, some part of the gene to take off it, some parts to add or an extremely sophisticated work to do to isolate it or put it together or identify the utility and prove the utility. And that will constitute work that will then entitle you to probably, – depending on the patent offices, but to try, at least, – to claim a patent on one of the two ways. One is going upstream like a product by process, the way in which you have obtained your product or the other end, i.e., you have identified a new utility for a naturally occuring product and in this case, if it is very strong and novel, corresponding to the basic rules in patenting, then you'd be entitled to go backwards in sort of defining a new utility that you have proven and identified, whereby you can claim the product that you get the utility from, whereas the patenting of the product itself per se would not be possible. That's what this explanation aims at.

Sahai: Which would be a very curious kettle of fish when you consider what happens with plant genes. All of these genes that are used are naturally occuring genes and to say that human intervention is an extraordinary step that makes it patentable strikes me as being extraordinarily curious because you have to do something even to recognize a naturally occuring gene which has happened all through classical plant breeding. There has been human intervention and it strikes me as being somewhat curious – to make that patentable is like saying, "I'll take this overhead projector out of this room where it's being used and it's natural and take it into that auditorium." That constitutes work and human intervention and make it patentable there.

St. Rémy: Yes, I mean, that's for patent offices to judge upon. And that's surely Craig Venter's and Bernadine Healy's hope.

Straus: I am in the fortunate position not to be a member of the Commission or somebody who shares lunches with the pressure groups and there is no danger that I would produce any leakage in the policy of the Commission. But with respect to the last discussion, I can read from the memo to the amended proposal which might enlighten a little bit the discussion. I quote, "With regard to the unpatentability of human bodies or parts of the human body the Commission wishes to make it quite clear in keeping with the discussions that have taken place within the Parliament that parts of the human body per se means parts of the human body as found inside the human body. It is important that this be spelled out so as to remove all possible ambiguity with respect to the position of certain products or parts of the human body which are already covered by patents granted in connection with the development of medicinal . . ." And there are two examples indicated here, for instance, a patent for a human lymphoblastoid cell line and the human hepatocyte culture process. So far, the Commission expressed its view in the memo, but of course the memo is not binding and therefore, I think this should be shifted into recitals.

When I asked for the floor I did't know that Dr. Rowland and the gentlemen from Heidelberg and also Prof. Vogel would address that issue, too. What bothers me and I think what should be addressed by all believers and critics in respect to the parts of the human body as, for instance, cell lines and the like, is that according to my modest opinion, as long as these parts of the human body can be commercialized, that means that you can buy a monoclonal or polyclonal antibody produced by such a cell line and you can buy such a cell line in order to produce such antibodies. The question is, how can you ethically justify, that you may commercialize antibodies, but you may not protect the invention how to produce them. This has two implications. The first is quite an ethical one. If you say there should be no patents on that; you say there should be free use for the followers. But if you don't grant any patent on that, everybody can step in in a later stage by a much cheaper procedure. This is one part of the story. And the second part of the story has been addressed by the colleague from Heidelberg. If, in the United States patents can be taken for that there will be an investment in these kind of products which are badly needed. And we will have to import them because in Germany, nobody will invest in that area. I mean, not science, but the companies need some protection in order to make those investments. This should be borne in mind when addressing this issue.

Folman: The question is now, what is going to happen on a global scale, really? If you've got a global effort to have a harmonization on commercial matters and also licensing, if you harmonize something in Europe and the U.S. will have different regulations and other countries, again, will have different ones, how do you foresee actually the future with regard to patenting things in biotechnology? Are we going to have a kind of controversy or a commercial war for many years to come? What actually is going to happen according to your view from the European Community?

St. Rémy: I'll have to answer this very briefly and I'm afraid I'll have to be bold. The GATT negotiations are not the beginning and the end of the world. Sovereignty in Europe, in the United States and in Japan still exists. GATT will not rule the world in all fields including research in biotechnology. That's point one. With due respect, I understand how important the GATT negociations are. What I mean is, not all legislation and all subjects worldwide should be commensurate with the GATT agreement, otherwise it's impossible to achieve. The second point is, I do think that one should not translate everything in terms of paranoid contention between different systems. One can, in many cases, go in a more flexible way and the patent lawyers and the patent managers in companies are very used to the ways in which one can turn around and understand rules. And again, back to Prof. Straus's intervention, I do think that the product by process or the product by utility can protect you against infringement. Obviously, selling, say, a cell line (a human cell line); would not infringe upon your patent because it would not be convered by a patent. Only it would be covered by a utility patent, i.e., if anybody having bought the cell line doesn't pay anything to you and produces your metabolite, that would be infringement because you would have covered by your patent the process to use the cell to produce the metabolite. And there would be an infringement there. And again, and thirdly, and also very briefly – it's a very long story – aside from the GATT but in coordination with the GATT on this particular subject there are, for the time being, very frequent meetings of two groups. One is the EC-US High Tech group – it goes to the level of the White House and the Commission presidency that we are trying to harmonize the right notably in the field of intellectual property and notably in biotechnology. It's been the item on the agenda for the last three meetings of this EC-US High Tech group. And secondly, at the same level, EPO and the American Patent Office and the Japanese Patent Office are also engaged in international, trilateral, talks that you might have heard of. So, we are trying to put everything under control by way of harmonization on the one hand and keeping the decision power that exists at the different levels free on the other hand. But again, briefly, there are discussion groups between the Community and the U.S. and Japan and there is also a group

– a very important, high level one, discussing th same matter between the EPO Washington office and Tokyo office. I think Mr. Gugerell could tell you a bit more about that if you wish so.

Beyreuther: I have only two short questions. The one question is, could a natural gene of a chimpanzee be patented even if it has exactly the same sequence as a corresponding human gene or at least if it would code for exactly the same product? That is the one question to understand the situation a little bit better. Now the second question is slightly more complex. Apparently, harmonizing plays an important role in the European Community. Harmonizing laws in gene technology a.s.o. But also ethics plays an important role. Now, is there any committee in the European Community harmonizing ethics in the Community? I think that would be important to have.

St. Rémy: Answer to question a) I think has been partly answered by Dr. Gugerell, i.e., it's not finalized, but the tendency for the last text would be to mean, in a way, that parts of the human body, in this case, a gene, is a part of the human body, but if taken from the human body, it is not – because if you find in a chimpanzee or a cat a gene that's identical to a human one, it is not a human one, it's a cat's gene. It's also in the memo that Dr. Straus has just read to you. And there is an ethics committee, indeed, that has been put together in November 1991 at the community level. It is an informal body but I can't see how one could have a formal body on such a point. But we have an agreement with a group of national experts. We have put together a so-called "Group of Advisors on Ethics in Biotechnology." That's all I can tell you and it's working hard but it is not in charge of drafting a legislation or anything of that kind. It can examine, and indeed, I can tell you it does examine any initiative at community level taken by the European institutions: a research program directive, a regulation, even a research project. It can examine anything that it chooses to examine and provide the Commission with its advice and if we do not follow the advice, we have to explain why and argue with the other bodies beyond the committee like the Parliament or the Council of Ministers. So, we think twice before choosing not to follow such advice. It is there, it is working and as I stated, we are currently expecting, on this particular item that we're discussing today, the formal advice of that European Committee on the patent proposal before the holiday.

Vogel: Now I wish to finish this discussion. Dr. Caskey could not come because of sudden illness. Now for the last minutes, Dr. Rowland was so nice as to tell us a little bit about the present state of the Human Gene Project because this has been more or less Dr. Caskey's topic and maybe at the end of this morning's meeting you may give us some information on this point.

Rowland: Since there will be no discussion, I guess I can say anything I please. I haven't done much science for about 30 years now, so this will not be quite from the point of the scientist, but I do work with universities and start-up companies so I have a fairly close relationship with them and their point of view. I think we've forgotten that in 1975 there was a meeting at Asilomar where the question was raised whether we should proceed at all with molecular biology and the decision was to go ahead, because while they felt it was a threat and could destroy us, they felt the benefits might be more than the threat that actually existed. In the human genome we should realize it's not just the human genome that's being studied, it's also yeast genome and I think the mouse genome. And so all of these studies are going on at the same time and to the extent that genes are conserved across these species they'll be used together to work out what is happening in our genome. But I think what't happening today, if you look at what a lot of science is concerned with, is the finding of proteins that have a physiological purpose. You then use that protein to try to find a drug. What we're finding today are not products but tools and much of what will come out of the Human Genome Project are these tools. This raises the issue that there is no uniform attitude about research throughout the world in patenting. The fact is that in the United States we have a very restrictive rule about doing research if there's a patent. You cannot do commercial research without infringing on someone's patent which is quite the contrary in the rest of the world. And so what we're going to see is an interesting situation where companies are being set up to use these tools to try to find drugs and would be subject to the laws in our country and yet, because we publish almost invariably either through the PCT application or subsequently – usually prior to that by various publications in journals – that the rest of the world will be free to use these tools and will try to develop drugs. In the harmonization of the laws, I think we have to look at how laws are interpreted and not just the statutory law because if we only look at the statutory laws we will ignore the various things that have developed over the years in understanding how we interpret them. I think the research exemption is very important. At least as far as I'm concerned, it needn't mean, that you in any way inhibit anyone from doing anything, but the question would then be, can we give our laws extraterritorial effects by refusing to allow you to import a tool into a country where there is a patent on the tool without having to pay a royalty. There is still some attitide about whether we should try to expand patent protection in the United States. From the patent side, I think, there will be continuous efforts to expand the reach of our patents in the United States in the biotechnology area and to try to protect the very large investment that is being made. And we do find that the foreign companies, Sandoz in particular, have been trying to buy into technology development in the United States and have met a very hostile reception. I don't know if you've read about the Scripps situation where Sandoz is willing to offer 350 million dollars over ten years for substantially all the rights in any inventions made at Scripps – not quite that far-reaching, but NIH threatened to pull any support out of Scripps if they went through with that contract. So, we still feel that we're the leader in the area of biotechnology. We are putting in a substantial proportion of the money into the Human Genome Project. A lot of that money will be to develop tools and not products. There will be some effort to protect investment, I think, and certainly I would suport it. We should provide some way to give incentives to develop these tools and then protect products made with those tools. I'm not sure that would be the scientist's point of view, but that's my point of view about the Human Genome Project.

Session 3

Legal Protection of Living Organisms from the Point of View of Scientists in Animal Breeding

H. Kräußlich

1. Sectors of Animal Breeding

Classifying domesticated animals into major groups of specific value to mankind and restricting that list to those species for which there are well established breeding industries, one comes up with the following categorisation:

a. Agricultural livestock
   - dairy, dual purpose und beef cattle
   - pigs
   - sheep and goats
   - poultry
   - rabbits
b. Horses
c. Fish kept for aquaculture
d. Laboratory animals
e. Pets and ornamental animals (poultry, fish)

This is not intended to be complete, but rather to identify the major groups of the animal breeding sector.

a) *Acricultural livestock* breeding is highly organised in developed countries and the breeding sector is divided into two quite distinct groups: cooperative breeding organizations producing purebred stock and breeding firms producing hybrids by crossing synthetic lines. Pure breeding is dominating in cattle and crossbreeding in poulty.
Breeding is very efficient in changing individual production traits by selection and systematic cross breeding. Milk yield per cow and year in Germany increased from approximately 1000 kg in the middle of the 19th century to an average of more than 5000 kg today. In some regions with intensive dairy production an average of 7000 kg, and herd averages over 10000 kg are registered. Similar alterations have been achieved in the ratio muscle to fat of pig-

carcasses, in the feed conversion rate of broiler and slaughter pigs and other traits. The extreme development of certain traits is the result of formulating distinct breeding aims according to the economic conditions and consumer preferences, and establishing of efficient breeding programmes.
The generation of transgenic animals has been successful in cattle, pigs, sheep, goats, poultry and rabbits (Brem 1993).

b) *Horse breeding* can be divided into four sectors: race horses, (thoroughbreds and trotters), riding horses, ponies and special breeds (e.g. pure bred Arabs). Modern animal breeding (herdbook breeding) has been initiated in England 200 years ago with the registration and standardisation of racing performance of individual horses and the consequent selection of superior horses for breeding. The economic motivation of horse breeders is difficult to evaluate as social factors (prestige) and tax incentives play a role in many countries. However, breeders respond to prices of foals and horses on auction sales like breeders of all sections of animal breeding. In the thoroughbred sector additional investment is attracted by the "syndication" of horses, especially stallions.
The generation of transgenic horses is not yet published.

c) *Fish* is an important foodstuff for human consumption, and for such important candidates in aquaculture like carps, salmon, and trout great expertise in fish raising and cultivation has been gained. Selective breeding schemes have been developed for these species during the last decades and the production of transgenic fish has been described for carp, catfish, loach medaka, salmon, tilapia, trout and zebrafish (Brem 1993).

d) *Laboratory animals* are used for basic research, toxity testing and other purpose. The first patented animal (US patent 4,736,866) is a laboratory mouse (named oncomouse because it carries activated human cancer genes). Laboratory mice are bred as inbred lines or outbred lines. The development of a new strain traditionally begins with a spontaneous mutation (e.g., the nude mouse) which is identified and bred directly or backcrossed to transfer the trait into an existing strain. Purchasers of laboratory mice normally receive the right to propagate the strain at will. Some strains are difficult to maintain in a pure form over longer periods so that breeding pairs must be purchased periodically from specialised breeding units from which Jackson Laboratory and Charles River Breeding Laboratories, USA, are the largest and best known. Private laboratories are supported in part by grant monies, and the sale of new strains is an additional source of income. Sometimes laboratories ask purchasers of newly developed strains *not* to breed it so that the laboratory could recover its investment. Besides private firms, publicly funded organisations including Universities maintain breeding units and develop strains as a research-support activity.

e) *Pets and ornamental animals* will not be discussed here, but future implications of animal patents to these sectors of animal breeding should be kept in mind.

2. Molecular Animal Breeding

New breeding techniques allow either isolation and direct transfer of the entire genome as a compact unit (e.g. nuclear transfer into enucleated oocytes in order to establish larger groups of genetically identical animals, clones) or direct transfer of gene constructs into the germ line of animals. The latter techniques give rise to transgenic lines of animals characterised by the stable transmission of the genetic modification.

In Biotechnology, Volume 2, Brem (1993) describes methods of gene transfer and its application. Gene transfer requires the combination of methods of different scientific disciplines as reproduction biology, embryology, molecular biology and animal breeding. Three strategies of gene transfer have been successfully developed: micro injection of gene constructs into fertilized oocytes, the use of embryonic stem cells and the use of retroviral vectors; alternative methods such as DNA-treatment of sperms have not yet been successful.

The estimated number of publications dealing with *transgenic mice* amounts to approximately 2000. The transgenic mouse is the most important experimental animal in medicine, in gene expression of higher organism and in developmental biology (Brem 1993). For patenting transgenic mouse lines first of all disease models are currently most important.

Research in transgenic farm animals is focused either on *traits of economic importance in agricultural animal breeding* or on drug farming. Traits of economic importance for farmers which have been studied are: growth and carcass composition, biochemical pathways, quality of products and disease resistance (Brem 1993). The practical application of transgenic animals is still a long way off. It will be necessary to improve the efficiency of transgenic techniques and to develop relevant gene constructs before the use of gene transfer in farm animal breeding including poultry and fish will lead to concrete and useful applications. It is to be expected, however, that with increasing knowledge coming from transgenic mice progress will be made also in farm animals. But, due to the longer generation intervals, expected progress remains slower.

The application of gene transfer in *drug farming* is more advanced. Many different proteins are currently produced by recombinant DNA technology involving the large scale expression of genes in bacteria, yeast and tissue cultures. But almost any protein the human body produces can also be made in farm animals, if their genes are programmed correctly. Thus, pharmaceuticals, diagnostics and food components can be produced by farm animals. The advantage of farm animals is that their products, such as milk, meat, blood etc. are free from toxic substances under normal hygienic conditions on the farm. Purification of the additional proteins produced in transgenics should be easier than in bacteria and yeast. Potential concerns of regulatory bodies may include the degree of genetic modification an organism undergoes, its potential threat to the environment, as well as the potential for impurities such as viruses. These problems are the same as in conventional animal production of food for human consumption.

For obvious reasons the use of the *mammary gland* of transgenic animals for the production of heterologous proteins is most attractive. The advantages are (Brem, 1993):

- In contrast to prokaryotik production systems mammalian cells are normally quite capable of carrying out the posttranslational modification of heterologous proteins.
- The recovery of the protein by milking avoids any adverse effect on the animals.
- The efficiency with which the mammary gland synthesizes proteins is enormous.
- Milk is a pure and hygienic product.

The mammary gland-specific expression of transgenes in farm animals has been reported for the rabbit, pig, sheep, goat and cattle. Besides the tissue specific expression in the mammary gland a high concentration of the transgene product (protein) in the milk is essential for economic gene farming. Wright et al. (1991) used the sheep β lactoglobulin gene fused to sequences of the human α_1 antitrypsin gene. Two of the generated transgenic sheep produced 1–5 g per liter while the third produced on average 35 grams of human α_1 antitrypsin/liter milk (50 % of total milk protein). Thus, overall yield of human α_1 antitrypsin might exceed 10 kg per sheep and lactation. The recovered α_1 antitrypsin shows similar bioactivity to the human plasma derived product. The level of human α_1 antitrypsin production in sheep exceeds those obtained in bacteria (15 % total cell protein), yeast (40 % soluble protein), and cultured mammalian cells. The highest concentration of transgenic products in milk has been reached in transgenic rabbits with a hybrid gene construct comprising bovine α_{s1}-casein gene regulatory sequences and the bovine prochymosin gene with up to 10 g prochymosin per liter (Brem and Hartl, 1991). Therefore, drug farming might make an important contribution towards the production of proteins which are of medical importance or which are required for other purposes.

Future development will mainly depend on the cost of bringing new drugs and biologicals to the market with different methods. Rajiv et al. (1993) compared the economics of production of recombinant tissue plasminogen activator (tPA) produced in Chinese hamster ovary cells and in Escherichia coli. They conclude for tPA and its subsequent-generation products that animal cells produce an apparently faithful copy of the endogenous protein but very large scale cultures are necessary to produce the amont of tPA that the market requires. E. coli is yielding large quantities of the product, but suffers from the need to operate an extraction and refolding process that is scarcely practical on industrial scale. Tufts University and Genzyme Corporation, USA, have generated transgenic goats in order to produce tPA in the milk. Future process economic studies comparing transgenic mammals, animal cells and bacterial fermentation will evaluate the systems and show the best integrated production system.

Besides gene farming the attempt to genetically modify organs and tissues of pigs to be used in *Xenotransplantations* will become one of the greatest challenges of gene transfer techniques in farm animals. The technical biological basis has not yet been worked out, but some first steps in this direction have already been made on a sound theoretical basis.

3. Legal Protection of Animals

3.1 Laboratory Animals

The first patented animal (US patent 4,736,866) is a laboratory mouse (named oncomouse because it carries activated human cancer genes). Mouse prices of transgenic lines such as the patented "onco mouse" (bred by Charles River Biotechnical services) and "knockout" mice with one or more genes "knocked out" by targeted gene transfer (e.g. bred by Gen Pharm) are at present 5 to 10 times higher than the price charged by nonprofit mouse breeders such as Jackson Laboratories of Bar Harbor, Maine. Gene Pharm additionally prohibited laboratories from breeding the mice, which forces researchers to pay for every mouse they use. For the "oncomouse", Du Pont came up with a pricing policy that contains a "reach-through" clause requiring anyone who develops a product through the use of the mouse, or any derivate strain, to pay royalties to Du Pont. As a result of this price policy many scientists try to breed their own oncomice in order to boycott Du Pont (Anderson, 1993). The National Academy of Science, USA, has held a meeting in March 1993 to discuss prices of transgenic mice and the way to combat the restriction of free flow of research materials among scientists. At this meeting Gen-Pharm announced that they allow researchers to breed as many as they want for an annual fee of $ 1000 and the initial purchase of a breeding pair (Anderson, 1993). Discussion about what an optimal solution of the marketing of transgenic lines might be is going on. The situation in Europe is somewhat different because of "research exemption" in Europe patent law which does not exist in US patent law. The transgenic research market, small as it is, is the first example for the practical problems which have to be solved in the different sections of animal breeding following the patenting of transgenic animals.

3.2 Possibilities to Patent Agricultural Livestock

Up to now conventionally bred animals and mainly biological processes have been summarily excluded from protection. Neither patents nor variety protection have been issued for farm animals. According to the "European Convention for the protection of animals kept for farming purposes" (proposal for an EC council directive VI/5143/89) farm animals are defined as follows: "All animals that are bred or kept for farming purposes ('animals bred or kept for the production of food, wool, skin or fur, or for other farming purposes, including animals produced as a result of genetic modifications or novel genetic combinations')". It is not yet clear whether the production of biomedical proteins in milk ("molecular farming" or "drug-farming") is interpreted as another farming purpose. The act of modifying an animal is not seen as breeding an animal.

A new situation has arisen during the review of the EC proposal for a Council Directive about the "Legal protection for biotechnological inventions". With the exception of livestock breeds and mainly biological processes the EC is likely to expand the reach of patents to nonbiological animal breeding processes and to the products arising from them. Even the discovery of the effect of existing biological material may become patentable, if this is a new discovery. Legal protection would cover the trade with breeding material, commercial animals and their products, such as meat, milk and eggs. According to the draft proposal, farmers would be exempted by a farmer's privilege to propagate genetic material for their own requirements without paying royalties.

It must be emphasised that the EC draft proposal is still being negotiated and *may come out differently*. COPA Working Group on Breeding Animals (Submission 30505 MR1) strongly objects to the patenting of whole animals after integrating one or a few genes into the genome complex. The newly integrated genes are only part of the genome, and the genome in its entirely should not be patentable, because the main part is the result of many years of conventional selection.

a) Examples for Gene Transfer Between Species That Could be Patented

Investigations to transfer the tithonin gene from the carp (Cyprinus capra) to cattle (Bos taurus) show that the modified tithonin gene confers longevity and continuing growth after sexual maturity to mice (Weiss, 1993). Viandegene Inc. in Chicago are making progress in this field in cattle and applied for an US patent (Bokanovsky and Bokanovsky, 1992), and the Kobe Soma Corp. is supporting a Japanese team. In order to insure a safe containment for transgenic animals it is tried to link apoptotic genes to the transgene target genes. Consequently, uncontrolled released transgenic animals would not survive because of apoptotic death (Weiss, 1993).

The generation of modified higher organisms by genetic engineering is viewed with concern by a high percentage of the population in some countries. This concern is multiplied by animal welfare groups, green parties and mass media. In a recent gene-ethics conference (Cassandra, 1993) a breeding moratorium was mooted after Bokanovsky suggested that apoptotic death in transgenic cattle would be both human and would provide tenderized meat. The ethical problems of patenting farm animals are discussed in some detail by Kräußlich (1992).

b) Examples for Biological Material that Could be Patented Within Species

Milk is probably the best known food with respect to its biosynthesis and composition and the chemical structure of its individual components. It is processed to provide a wide diversity of food products which comprise more than

25 % of total consumed dietary protein in the industrialized world. Improvement of milk protein quality may be achieved by altering physico-chemical properties of individual milk proteins, in order to enhance their functional characteristics; by modifying the relative proportions of individual components, to improve the heat stability of milk (e.g., by reducing casein micelle size); or by adjusting the protein composition for nutritional requirements (e.g., "humanized", milk elimination of β-lg.).

Characterisation of polymorphism of the relevant genes within species provides information necessary for developing *allele specific typing procedures.* Relying on polymerase chain reaction this procedure has proven to be a potential tool for selection in farm animals.

Martin and Grosclaude (1993) investigated the goat α_{s1}-casein gene polymorphism and this is an excellent model to study the interactions between caseins and other components of milk, in order to understand how modifications in individual components will influence the processing behaviour of milk and the organoleptic characteristics of the final dairy product. There is no doubt that in future the combination of new (gene targeting) and established (selection) genetic approaches allows more tailored genetic improvement which may fulfil all requirements for patenting of animals.

3.3 Specific Problems in Patenting Agricultural Livestock

a) Farmers' Exemption

Transgenic cattle with the tithonin gene would produce more, leaner meat from less feed. The patent would extend the inventor's monopoly to all the calves bearing the tithonin transgene. Any farmer who bought a tithonin-transgenic calf could be entitled to rear or fatten it and sell it for meat. Assuming farmers' exemption will be accepted the farmer could in addition legally breed the animal and raise its offspring on his farm without the permission of the patent holder (some of those offspring will receive the tithonin transgene). Farmers' exemption is the rule for plants covered by plant variety protection. A farmer who bought protected seed has the right to replant seeds harvested from the protected variety on his own farm.

In my opinion, there exists no realistic way to bar farmers from breeding several generations of a protected nonhybrid strain of farm animals on their own farm because it seems impossible to control all farms.

However, sales of transgenic breeding animals, semen or embryos from transgenic animals can be controlled by breed societies and it would not require additional organisations to collect royalties for patents.

b) Exemption for Experimental Breeding and Research

Without exemption a holder of an animal patent would have the right to limit the use of patented animals for experimental breeding and research. If experimentation is permissible only to satisfy academic curiosity but not for commercial purposes, the progress of breeding will be slowed down. Animal breeding would be best served if all laboratories, commercial and academic, had the right to experiment more or less freely with patented organisms. That freedom would increase the incentives of follow up research. In the example of the tithonin transgenic cattle Japanese researchers used a modified version of the tithonin gene by cloning tithonin from Koi carp that live in the thermal springs on Mount Asama. The Koi-tithonin transgenic mice terminate growth at a weight of 1,5 times normal body weight and appear to be healthy in every other respect. Assumed, Bokanovsky and Bokanovsky had been granted the first patent under a narrow research exemption the research of the Japanese group would have not been possible without prior permission of the original patent holder.

c) Acceptance of Legal Protection by Animal Breeders

In the absence of practical experience with patents, animal breeders fear the increasing royalty costs of future innovations and the replacement of existing breeding organisations by commercial firms holding patents. On the other hand, the need for investment protection for innovative animal breeding is recognised by breeding organisations.

In parallel to plant breeding, the breeding organisations in the European Community are demanding a special animal variety right to regulate biological breeding processes and products of animal breeding that are more uniform than conventional breeding results, but less uniform than is required for patentability. Equally, certified animal varieties would require a lower degree of novelty than would patents.

Considering the problems of control on farm as mentioned under a) above, such an animal variety right could serve as a bridge between the requirements on farms and in breeding organisations on the one hand, and the integration of patented animal material on the other.

Lines or animal varieties that combine several novel genes will almost certainly be economically preferable to those with only one new gene. Inventors or companies could licence new genes to animal breeding organisations, who can combine selected genes in transgenic animals that will appeal to particular markets. In this scenario a farmer could buy animals that have one company's disease resistance gene, another's growth gene and a third's package of metabolic pathway genes.

d) Influence on Farm Structure

Animals breeders fear that patenting transgenic livestock will throw the small breeders out of business. Proponents of patenting argue that improved transgenic strains are likely to be invented and used in any case, even in the absence of patents. However, in the absence of royalties for patents, inventors are likely to use contracts as an alternative means of recovering costs. Such contracts could be administered most efficiently with large farms, meaning that small farms might not get access to new technologies without patenting of farm animals. Royalties only will be counterproductive for the small farmer if the costs of collection are too high and absorb much of the productive enhancement of the transgenic animals. Structure and organisation of breeding programmes in regions with small farms are therefore very important for small breeders.

e) International Exchange of Pedigree Animals, Semen and Embryos

International exchange of pedigree animals, frozen semen and embryos is the rule in breeding of farm animals. Animal breeders fear that patents may be used to exclude some breeders or breeding societies from the use of the invention. Consequently, the genetic exchange between populations would be restricted and in extreme cases breeding sectors might be monopolized.

f) Traditional Purebred Producers Versus Large Breeding Companies

It seems highly unlikely that herdbook organisations will be able to carry out all the biotechnical research which is necessary for patenting animals. The research will be done by universities, government funded institutes, and private companies, which will be patent holders. Only large breeding companies like American Breeders Service, USA, have the potential to develop their own biotechnological innovations and to introduce the patented genes into animals of commercial pure breed lines for sale to farmers. If transgenic methods become routine traditional breeders organizations can license patents from the holder and introduce the patented genes into their commercial purebred lines or animal varieties for sale to the farmer. Incorporation of biotechnological inventions without major changes within existing breeding structures and practise my be possible in future.

4. Acceptance of Animal Patents by the Consumers and Breeding Goals in Farm Animals

In spring of 1991, a public inquiry was held about biotechnology in the EC: Eurobarometer 35.1 [INRA, 1991]). About half of the EC population feel that their quality of life will improve as a result of applications of biotechnology or gene technology. The results show striking differences between the various countries, in Denmark and Germany people have the lowest expectations of gene technology. On the question what research areas are worth the effort, biotechnological research with farm animals came out worst and biotechnological research aiming at obtaining human medicines came out very positive.

Comsumer organisations and environmental organisations are considered the most reliable sources of information about risks of biotechnology by the majority of populations. The application of biotechnological methods in animal breeding and in food production has at least in Germany an intensive and frequently adverse coverage in the media (for instance "Gen-Fraß", Der Spiegel, Nr. 15/47.

Jahrgang, 12. April 1993). Regarding gene transfer the main reason may be that this field combines two very sensitive issues to which the public reacts with distrust if not aversion, i.e., genetic engineering and animal experiments.

A large number of societal organisations have grouped around the European Parliament, examples being the cooperating farmer's organisations (COPA), the cooperating animal protection groups (Eurogroup for Animal Welfare), cooperating environmental organizations (Friends of the Earth), the European Environmental Bureau, the Greenpeace International EC, Unit, the International Organization of Consumers Unions (IOCU). Eurogroup for Animal Welfare and Friends of the Earth have principal ethical objections against genetic modifications of animals (for instance: transgression of boundaries of species) and see problems because of the risks for animal welfare. In general, lobby organizations tend do oppose patenting of animals rather than seeing advantages.

The Dilemma of animal breeding is that breeding goals are determined to a great deal by consumer demands. Consumer demands may change rapidly but the change of animals by conventional breeding methods is slow. Consumer demands and public opinion must not be in agreement. It is not an unlikely scenario for the future that public opinion strongly opposes transgenic livestock in the own country and products of transgenic livestock produced in distant regions conquer the market due to efficient marketing.

5. Summary

Traditionally bred animals have been summarily excluded from legal protection (patents or variety protection). Patenting of animals in which a useful gene is transferred from another species has become possible (US patent Nr. 4,736,866; 1988), and the EC is likely to expand the reach of patents to non-biological animal breeding processes and to the products arising from them.

During the past decade methods have been developed which, unlike conventional breeding techniques, allow the genetic composition of organisms to be modified directly and selectively rather than indirectly by estimating the breeding values on the basis of phenotypes in order to establish a ranking order for selection. In principle, these new techniques of genetic manipulation can be categorized into those which treat the entire genomes as a compact unit (e.g. nuclear transfer for establishing a group of genetically identical animals), and those in which individual genes are manipulated (e.g. gene transfer). The application of gene transfer techniques has furnished us with new insights in developmental biology and the principles underlying tissue specific gene expression. It has also provided oncologists, immunologists, and medical geneticists with a plethora of important and detailed information. In addition, gene transfer also allows the development of new production system for pharmaceutically important proteins (drug farming). As far as agricultural animal production is concerned, gene transfer appears to be a promising technique for improving disease resistance, the performance and the quality of animal products by modifying, for example metabolic pathways, hormone status, termination of growth and longevity.

Domesticated animals can be classified into major groups of specific use to mankind (e.g. agricultural livestock, horses, pets, laboratory animals, fish, ornamental animals). Gene transfer techniques are used in mice (disease models), agricultural livestock (drug farming and genetic improvement of economically important traits in cattle, sheep, goats, pigs, rabbits), poultry (chicken) and fish (aquaculture). Disease models and drug farming are at present the most important applications for patenting. But animals kept for farming purposes and ornamental animals may become even more important in the long run.

Adaptation of patent right to the specific structural and organisational conditions of animal breeding seems necessary. The creation of a special breeding right for animals (comparable to special plant variety rights), introduction of farmers' exemption and exemption for experimental breeding and research into patent right are at present under discussion.

6. References

Anderson C (1993) Researchers win decision on knockout mouse pricing. Science 260: 23–24

Bokanovsky BS, Bokanovsky SB (1992) Illinois patent file, no ZO 10493 A

Brem G (1993) Transgenic Animals in: Biotechnology, 2nd ed, Vol 2, Edit. Rehm HJ, Reed G. VCH, Weinheim, New York, Basel, Cambridge, 745–832

Brem G, Hartl P (1991) High-level-expression of prochymosin in the milk of transgenic rabbits, in: Frontiers of Biotechnology in Agriculture. Conference 1–4 August, Sea of Galilee, Israel

Cassandra S (1993) in Genethics and Transgenics (ed Orestes, AG). Academia, 895–987

INRA (1991) L'opinion de Europeens concernant la biotechnologie en 1991, INRA (Europe)-European Coordination Office SA/NV, Brussels

Kräußlich H (1993) Gewerblicher Rechtsschutz und Tierschutzrechtliche und ethische Probleme. Dtsch tierärztl Wschr 100:79–82

Martin P, Grosclaude F (1993) Improvement of milk protein quality by gene technology. Livestock Production Science 35:95–115

Rajiv VD, Cartwright T, Rosen CG (1993) Process economics of animal cell and bacterial fermentations: A case study analysis of tissue plasminogen activator (Review). Biotechnology 11:349–357

Weiss RA (1993) Dorian Gray mice. Nature 362:411

Wright G, Carver A, Cotton D, Reeves D, Scott A, Simons P, Wilmut J, Garner J, Colman A (1991) High level expression of active human alpha-1-antitrypsin in the milk of transgenic sheep. Biotechnology 9:830–834

Legal Protection of Living Organisms from the Point of View of Scientists in Plant Breeding

I. Sela

Breeders produce a new line, or variety, or a hybrid plant of commercial value following an elaborated and long procedure. This involves all, or some, of the following steps: the spotting of the desired trait in Nature, attempts to introduce it into a plant which can be crossed with the plant species to be improved, and the actual introduction of the desired trait into the commercial variety. The production of a stable commercial variety takes many years of crossing, backcrossing (and/or selfing) and selection after each step. The final plant "product" is then protected legally by "Breeders' Rights". The plant is described phenotypically, and any deviation from the initial phenotypic determination in a plant produced by another breeder, will render this plant unprotected. Thus it is easy to circumvent Breeding Rights protection. As a safeguard, most breeders try to produce heterozygous plants carrying the trait of interest. The breeders do not disclose the identity of the parent plants, and since the offsprings of the hybrid will segregate, they are much less usable for the farmer, who comes back to the breeder for a new supply of seeds. However, it is relatively easy to "break" a hybrid and produce a variety, or another hybrid. Thus, practically, a real good hybrid would not be protected long.

Recent developments in molecular biology expands the spectrum of the gene pool that potentially can be introduced into a plant variety, and provide tools to identify a genotype rather than a phenotype. The ability to produce transgenic plants enables the insertion, into a commercial plant variety, of genetic traits from other species or genera, or even from sources other than the plant genome. For instance, we were able to produce a tobacco plant which carries the gene for human β-interferon, and the expressed protein, made in the plant, is active in human cells. We have also transformed plants with viral genes that render them protected from viruses. Transgenic plants can be patented, and this is the first example where a DNA sequence in a plant, rather than a phenotype, is protected. Analysis of the situation raises a few problems that are discussed below.

The molecular technology, when integrated into the breeding program, enables breeders to shorten the breeding program considerably. The genetic trait which is being bred can be (in many cases) labelled molecularly, thus a molecular selection may soon replace phenotypic selection. However, while the breeder may soon rely on molecular markers in producing a new variety, legally he can not yet protect it by the same molecular criteria. This also arises a number of problems that will be discussed.

In the context of patenting of living organisms, plants are different from human beings and animals in the sense that individual members of large plant populations may be absolutely identical, each individuum having the same genetic makeup. Many Agricultural plant "varieties" are, in fact, an outcome of clonal selections and are propagated vegetatively (asexually) from a single selected "mother" plant. Thereby they comprise a large population of "identical twins", carrying the exact same genetic composition. When two plant varieties (or even different species) are crossed, an offspring plant can be selected for a specific trait and submitted to several cycles of backcrossing with one of its parents. At every generation, a progeny plant is selected, and is being backcrossed with the same parent. After several backcrossings, a "near isogenic" line (to the pertinent parent plant) is thus produced, which is identical to that parent but carries a new genetic trait with a minimum stretch of unrelated DNA. Hybrids, F1, plants also compromise a population of, commercially important, identical individual plants. These hybrid seeds are better protected than newly developed varieties, since in addition to being protected by Breeders' Rights, their parents are not disclosed. Hybrid plants are often produced from parents belonging to two near isogenic lines, which are, by themselves, almost identical.

While in the case of human beings the object of patenting is a particular gene, plant-related legal protection is directed to the organism as a whole. In this respect, establishing a definitive way of identification is essential. Phenotypic identification is, in my opinion, insufficient. One plant may be placed under certain conditions, and another, identical one, placed under different conditions, and the two "identical twins" will react differently. One may work hard and develop a certain plant with a very important commercial trait, and another party may just alter a single, minor, phenotypic characteristic in order to circumvent its legal protection. Through understanding of the processes, the legal community should develop a new approach to protect breeders, which will involve the genetic identification of the trait itself (protecting the gene), and the addition of molecular identification marker systems (legally equivalent to human fingerprints) of hybrid plants. Methods such as DNA fingerprinting, RFLP, RAPD, target-specific PCR, etc., should become legally admissible.

New lines of desirable plants (and to a much lesser extent also animals) may be obtained through the construction of transgenic plants. For instance, we have constructed transgenic plants that express and produce active human interferon, and other plants (some have been patented), which have become virus-resistant. Only a small proportion of the transgenic plants meet the pre-set standards. Hence, populations of transgenic plants should be screened, the desired individuals selected and brought to an homozygous stage, before a stably transformed plant is obtained. In this regard, one of the major, yet unanswered, questions relates to the fact that a certain characteristic is observed in only a few of the transgenic plants, even though they carry and express the same genetic construct. One possible explanation is that a slight DNA modification accidently occurred during the long process of transformation: a minor deletion or insertion, recombination, etc. Hence, the population of transgenic plants is not homogenous anymore with respect to the inserted DNA. If this is correct, the inserted piece of DNA in the plant will not match correctly the the sequences given to the patenting office, and deposited as plasmids at the pertinent office. Are such plants still legally protected?

It is my opinion, that molecular criteria should be developed, understood and accepted also by our colleagues in the legal profession. The technology is available, it should only be channeled and developed to the point where it answers specific legal questions. Every case should be studied for itself, requiring understanding of molecular genetics on the part of patent officers and the Courts. Phenotypic criteria solely should not bear as much weight in decision-making, as they currently do. On the other hand, the degree of DNA homology which will still be legally considered as "identical" to a given sequence should be determined.

"Patenting" of Living Organisms – Patents and Plant Breeders' Rights – From the Point of View of Plant Breeders

P. Lange

Breeding new plant varieties has always been a very time-consuming and expensive undertaking. This has not been changed by the use of the most modern technologies including cell biology and genetic engineering. While improvements in yield and nutritional quality were once the primary objectives of breeders, today they increasingly strive for resistance against diseases and pests, more efficient use of plant nutrients or properties for alternative uses. Even when making use of the most modern technologies (for many plant breeders the state of the art ranges from the use of cell biology methods to the transfer of desired qualities through genetic engineering) the costs required by plant breeding firms for the development of new plant varieties are at least four times higher than their costs of biotechnology. At present, the employment of these modern technlogies creates at best new variability in the form of individual plant properties which must be transformed, introduced and fixed into the genome of a new plant variety using conventional plant breeding methods – without impairing its other important agronomic properties. A plant breeding company such as KWS spends approximately 15 % of its turnover for research and development – a very high percentage compared to other research-intensive enterprises.

This makes it necessary for the plant breeder to optimally protect his intellectual property against imitation or theft in order to not endanger his basis of the return of investment. I do not have to explain to persons in this branch that this necessity is increased by the fact that the material to be protected is biological material which, in principle, manifests the ability to reproduce on its own. It is not dependent on "technical production" making use for a complicated manufacturing process – one need simply to exploit the natural reproductive abilities of plants.

In this respect one should assume that the plant breeder strives for the strongest possible protection of his intellectual property in the form of new plant varieties characterized as a unique combination of genes *or* as specific genetically manipulated constructs. But this is only one side of the coin.

The other side is the fact that even modern *breeding* – in contrast to a technical invention – *is always based on existing material* and therefore requires the broadest and most liberal use of plant materials possible [1]. This means that the breeder must be able to independently make use of protected plant variety materials owned by a competitor for his own further breeding purposes and to market new varieties resulting from this use; for example by crossing this material and thus developing a "mixed" new plant variety.

This also shows that good legal protection for a plant breeder always must take the two opposite interests:

- optimal protection against imitations and
- the most liberal possible use of materials, including protected plant varieties of competitors

into consideration. This is ensured in the area of plant breeders' rights by the principle of "breeders' privilege" – a central part of the UPOV Convention – but would not be ensured in all cases by the research exemption of patent law (which I think is the most consequential reason for banning patents for plant varieties per se in accordance to article 53b of the EPC).

This puts me to a significant difference between patent law and plant breeders' rights, the two main types of protection which come into consideration for the intellectual property of the plant breeder.

I think I should indicate some (not all) other important differences between these two forms of protection in order to explain the attitudes of the breeders to this complex problem (Table 1 [2]).

Table 1

	Patent	**Plant Breeders' Rights**
Criteria	"enabling disclosure of the **novel invention**" (technical teaching . . .)	(also discovery) **plant variety** (no teaching . . . the existence of the p.v. is sufficient)
	non-obviousness (inventive step)	**distinction** (small steps sufficient)
	industrial applicability	–
	–	**uniformity**
	–	**stability** (maintenance breeding)
Scope	. . . **using** the invention	. . . special acts (esp. selling) in respect of: – propagating material – harvested material – products directly obtained
	– research exemption	– **breeders' privilege**

This diagram shows on one side already the considerable differences in relation to the object of protection – *an invention* (technical teaching) or *a plant variety* per se (which must be distinct, uniform and stable), whereby the terms "uniformity" and "stability" are particularly conform to biological materials. In addition, the term "distinctness" also allows the protection of new plant varieties which are relatively close to varieties already under protection, which would be impossible under patent right because of the term "non-obviousness" or inventive step.

But the most important aspect is the difference described above between "research exemption" according to patent law and "breeders' privilege" according to plant breeders' rights. In the opinion of plant breeders, the patenting of plant varieties would in most of the cases prevent the free use of such protected genomes necessary for their work. European legislators apparently also recognized this since article 53b of the EPC stipulates that:

> "European patents shall not be granted in respect of plant varieties . . . or essentially biological processes for the production of plants . . . – with the exception of microbiological processes and the products thereof."

This rule, which in my opinion should not be interpreted narrowly [3] but, as usual for such delimitation standards, according to the aim and object of the standard [4] (this is not an exceptional provision), therefore allows patent protection for microbiological inventions and does also not generally exclude plants from patent protection. This provision prohibits only the patenting of plants and their propagating material in the genetically fixed form of a plant variety.

So, provided that general patent conditions for such an invention are met, patent protection is not excluded and should not be excluded. One example of things suitable for patent protection is: genetically manipulated constructs in plant material coding for specific proteins – such as virus resistance. Such patent claims regularly encompass plant material which is distinguished by the fact that it contains this specific gene or genetic construction which is revealed in the plant material. If such claims did not go as far as to cover plant material, the necessary return of investment from these inventions would not be secured for a very significant future range of application. Thus we can say that in the perspective of plant breeders such protection is not only acceptable, but desirable.

On the other hand, however, patent offices and legislation must ensure that the prohibition of patents for plant varieties per se quoted above is not undermined for the reasons already mentioned. This could happen, for example, if patent claims are formulated in such a way that they avoid the term "plant variety" but in reality encompass such material.

Therefore it has to be guaranteed that plant breeders are able to make free use of plant material in the form of complex genotypes for their breeding purposes provided that this no longer contains the genetic information patented. Our goal is to retain the genetic pool in its entire variability for breeding – not to use individual patented genes without compensation. When interpreting or applying article 53b of the EPC consideration should be taken that in such cases the core of the invention is the new genetically manipulated construction, not the unique combination of the plant's genome.

These fine differences may seem highly sophisticated to you, have little link to practice or at least appear difficult to apply. You may also ask how the exact border lines can be drawn – at least according to European law – between inventions which can be patented and plant breeding developments suitable for plant breeders' rights.

This question is justified enough by the different qualities of the protective systems described above, but in particular because products of development will in fact more and more exist in the very near future which contain both elements: for example plant material protected by plant breeders' rights with inserted patented genes.

At this point I have to mention that the revised 1991 UPOV Convention [5] supplies a solution for this very case. It contains the concept of essentially derived varieties which can only be commercialised with the permission of the owner of the initial variety (I would call this concept "plant breeders' rights dependency").

The revision of the UPOV Convention was necessary essentially for three reasons:

1. New possibilities for the use of modern plant varieties (for example in the form of end products, ingredients, etc.) and technically expanded possibilities for reproduction (rapid and comprehensive reproduction, for example using in vitro techniques) require an extension of the range of protection from reproductive material – over harvested crops to the products manufactured from them.

2. Misuse of the breeders' privilege and the rule of distinctness in the form of plagiarized plant varieties (such as spontaneous mutations of ornamental plants).

3. Protection gaps between patent protection and plant breeders' rights protection – for example in the case of "genetically modified new plant varieties".

The third reason is particularly relevant to the circumstances being discussed here. As I have already mentioned, when fulfilling the usual requirements for patents, genetic engineering procedures and their products are capable of being patented and many such patents have already been granted (for example precisely described genes or genetic constructs which cause a certain resistant property in plant material). This means that the owner of the patents for such genetic constructions can formulate and pursue his patent claims to the extent of plant material, without having to ask for permission of the breeder of the relevant genotype (or protected variety) let alone granting him the right to compensation for use of the results of his work as a breeder. This is a consequence of breeders' privilege described above which itself is not being put in question.

I also do not question the justification of protection from cost-free use of patented genetic constructions. But there was a distinct gap in protection to the disadvantage of the plant breeder.

The 1991 revision of the UPOV Convention did away with this protective gap an, in the viewpoint of the breeder, carefully balanced the ambivalent interest I cited above:

The Convention introduced the concept of "essentially derived plant varieties" [6]. The major contents of this innovative regulation can be summarized as follows: If the normal protection requirements (novelty, distinctness, uniformity and stability) are fulfilled, the breeder of an essentially derived variety is granted plant breeders' rights and may thus assert all rights resulting from variety protection against any third parties. However, the *marketing* of the derived variety requires the permission of the breeder of the initial variety, from which it was essentially derived. The result is that the "breeder" who desires to exploit an essentially derived variety requires the permission of the owner of the initial variety. This permission can be completely rejected or, what will happen in most cases, require the fulfilment of certain conditions, for example the payment of a licence fee.

This means that the one-sided patent dependency [7] – which existed up to now for the use of patented genetically manipulated products or processes – has been balanced out by the so-called plant breeders' rights dependency to the advantage of the work of the *original* breeder. For this purpose the core of the old UPOV Convention – breeders' privilege – was maintained. The essence of this new concept does not cover classic breeding procedures so that crossing another protected variety in existing variety material and the ensuing selection in the progeny do not lead to dependency. This is not a case of an *essential* derivation because no predominant derivation happens, not to mention that in most cases the required genetic conformity would not exist.

So, what is an "essentially derived variety"? Regarding the initial variety, the following three prerequisites must be fulfilled in addition to the existence of the essential attributes of the original variety:

a) clear distinctness
b) essential derivation
c) genetic conformity

If one of these requirements is not fulfilled, it is not a case of an "essentially derived variety".

Regarding a): The clear distinctness is judged as before based on expressed characteristics of the derived variety.

Regarding b): The breeder of the derived variety must have derived his breeding product essentially from the initial variety. If he did not use material from the initial variety in the breeding process, then no derivation in this sense has taken place.

Regarding c): In addition, the derived variety must genetically conform to the original variety. A classic case of such genetic conformity is when an additional gene is implanted in an existing variety by genetic transformation.

There will certainly be some problems in the implementation of this new principle. The revised UPOV Convention only formulates the principles of the new concept. But it offers a basis for a bridge between both protective systems in the interests of appropriate legal protection adapted to the biological matter in the area of plant breeding encompassing biotechnological developments. This legal framework is intended to offer a chance for the chiefly small-scale plant breeding sector – particularly existing in Germany – to continue receiving compensation for their considerable breeding and research expenses.

Finally, please allow to point out a critical aspect of the future: The latest developments in the field of intellectual legal protection have intensified the mutual understanding for the need of adapted protection in the various fields. The strengthened plant breeders' rights system will play an increasingly important role in the breeding industry as well as in the field of commercial plant biotechnology. But patents will also become more important. In addition, we can be sure that the entire breeding sector will probably be confronted more and more with different and various protective rights. This is already manifest in the mutual dependency resulting from cases of patented genetic constructs within a protected plant variety. For example it was recently pointed out that during the development of a single transgenic plant variety with resistance to a certain insect, eight different types of interdependent protective rights, probably involving different parties would be applicable (Table 2 [8]). Therefore stronger intellectual property rights for plant related inventions will better be able to protect real innovations in plant breeding. However, before a product is commercialised the proprietary rights of others will have to be considered. This may result in a complicated network of proprietary rights for some products.

Table 2. The Development of a transgenic plant variety resistant to insect attack leading to multiple rights

Subject	**Components**	**Example**	**IPR**
Plant Variety	Germplasm	Protected Variety, e.g. potato	PVR
Selectable marker gene	promoter coding sequence	35S neo	patent patent
Trait	promoter	TR	patent
	coding sequence	bt2	patent pending
Transformation Technology	Ti-plasmid	pGV2260	patent
Gene Expression Technology	Transcription Initiation	viral leader	patent pending
	Translation Initiation	Joshi	–
	Codon usage	AT → GC	patent pending
	Number of different IPR		8

Sehgal S, Van Rompaey J, IPR Complexities in the Global Seed Industry: "Seed World", May 1992, p. 24–25

I hope that such a complicated system of rights will not obstruct independent research and development but will lead instead to the competitive introduction of these products onto the market, to the advantage of the intellectual owner.

In the long run such products will only be successful on the market if their advantages are worth the cost involved. In this respect I am optimistic that the powers of the market will help to regulate such complex problems.

References

1. See also Lange P, GRUR INT 1985, 88
2. Lange P, GRUR INT 1993, 137–143 (139)
3. Dissenting opinion: Straus, GRUR 1992, 262; Moufang, Genetische Erfindungen im gewerblichen Rechtsschutz 1988, 190
4. See Benkard, Bruchhausen, PatentG, 8. Auflage 1988, 202
5. The new UPOV Convention is published in GRUR INT 1991, 538
6. See in detail Lange P, GRUR INT 1993, 137–143
7. See also Böhringer, comment in a public hearing, German Bundestag – Rechtsausschuß – remark 2, page 189 (201)
8. Sehgal S, Van Rompacy J, "Seed World" May 1992, 24–25

Genetic Diversity and the Indian Seed Industry

M.S. Swaminathan

I am grateful for this opportunity to speak on a problem of vital importance to the future of our agriculture and to the welfare of our farmers. The title for the discussion is "Genetic Diversity and the Indian Seed Industry". I would at the outset like to spend a few minutes on the basics.

Genetic diversity is the very backbone of sustainable advances in agriculture because without diversity we cannot make selection, or recombination. All progress in plant and animal breeding is associated with the diversity characteristic of the living world. Diversity in the physical environment helps to preserve and multiply genetic diversity, and this is why we say "variety is the spice of life". Forty years ago, i.e. in 1953, the molecular basis of how such variation occurs and is multiplied was discovered through the elucidation of the double helix structure of the DNA (Deoxyribose Nucleic Acid) molecule by Drs. James Watson and Francis Crick. And now forty years later we are fast mastering methods of manipulating variability at the molecular level. Thus, in addition to the normal methods of sexual hybridisation we can now resort to para-sexual methods of mixing genes so that altogether novel genetic combinations can be produced. Modern bio-technology, particularly recombinant DNA technology, has enhanced the value of variability because we can move genes across sexual barriers, and create transgenic plants, animals and microorganisms.

The ancient civilizations of Middle East, India, China and Central America were essentially based on the domestication of wild cereals. Barley and wheat were cultivated in Iraq about 10,000 B.C., rice and millet in India and China around 6000 B.C., and maize in Central America at about the same time. The remarkable botanical knowledge of huntergatherers is now well recognised. Unlike the situation today where global food security depends hardly on 15 major plant species, the food plants of North American Indians consisted of 1112 species belonging to 444 genera and 120 families. Right from the domestication of plants, farm women have used naturally occurring genetic diversity as a method of crop improvement. In fact, the very process of seed selection and preservation led to selection not only for higher yield but also for stability of performance through resistance to pests and pathogens. For example, even now in many parts of Indonesia a woman who harvests paddy often carries two bags. In one bag she keeps the panicle which will provide seeds for the next crop. The other bag is for grains. If we visit the Kulu Valley during *Dussehra*, some good looking maize cobs can be found hanging on the door. These are the seed cobs for the next sowing. Such cobs are the best the women select out of the harvested maize cobs. In other words, farm women and men have over a long period of time, tried to identify strains possessing resistance or tolerance to both biotic and abiotic stresses. Traditionally, it has been women who have been the seed selectors and conservers and have practised what we now call the mass pedigree system of selection by which they have been able to advance, sometimes by one percent per year, the yield potential of the crop. So we owe to the farm and tribal women and men of the last ten thousand years the enormous amount of variability we observe within a species.

The Global Convention on Biodiversity signed by over 150 nations at Rio De Janeiro in June, 1992 recognises diversity at three levels: first at the ecosystem level, for example the Himalayan ecosystem or the desert ecosystem or a coastal ecosystem. Second, diversity can be measured at the level of the species. Only about 1.5 million species have been described so far in literature by biosystematists. Some experts feel that there may be fifty million species and others even 80 million species. But the fact remains that the 1.5 million is just the tip of the iceberg. There are many more species awaiting discovery and description, particularly at the level of invertebrates and microorganisms. Unfortunately species extinction is taking place at a fast pace largely due to the destruction of habitats rich in biodiversity. There is thus an urgent need for the conversation of species and for the restoration of degraded habitats. We also need more biosystematists who could help describe the non-described species.

Genetic diversity is the hallmark of life on earth. Such diversity, in turn owes much to cultural diversity and pluralism in relation to protection, preservation and distribution. Every country and every language have sayings to stress that "we reap as we sow." During the process of domestication, emphasis grew more and more on yield and economic plants started depending on human beings for both protection and propagation. In other words, in nature, they had to fend for themselves; and hence had to possess the ability for self-protection and self-propagation. But under cultivation, human beings protected the plants and propagated them. That is how the seed industry started growing. For the purpose of the seed industry, we can broadly categorize crops into three major groups. First, in self-pollinated cereals like rice or wheat, farmers can keep their own seeds. Farmers tend to change the seeds when new varieties become available or when the purity of a variety goes down in a visible manner. The second group are cross-pollinated plants where special methods of seed production will have to be adopted. However, F1 hybrids are now being grown extensively in China even in a self-pollinated plant like rice. In such cases seeds will have to be purchased for every crop. There is a distinction between hybridisation which is inherent in breeding and the use of F1 hybrids as seed material for taking advantage of hybrid vigour or heterosis. For becoming commercially viable, F_1 hybrids should give 20 to 30 percent more yield than the parents. In the United States, hybrids revolutionised corn or maize production in the thirties. The exploitation of hybrid vigour has been in progress for a long period of time unconsciously or consciously. The mule among animals is always cited as an early example of the use of hybrid vigour. Chinese scientists led by Dr. Yuan Long Ping developed a methodology for producing on a large scale F1 hybrids in rice. The commercial exploitation of hybrid vigour depends much on the dynamism and state of development of the seed industry. In our country hybrid cotton was produced by hand-emasculation and pollination largely by tribal and rural women and men. They were able to emasculate and then pollinate with care. Thus, India became the first to start a hybrid cotton industry. Our rich human resource in villages places us at a comparative advantage in the exploitation of hybrid vigour in seed propagated plants, either through a male sterile-restorer mechanism or a hand emasculation cum pollination technique.

The third group is vegetatively propagated. Several horticultural plants like potato are propagated by tubers. In the case of mango or citrus, bud material or grafted plants can be used to raise orchards. Grafting can be used for rejuvenating old orchards. It can also be used in plantation crops like coffee, where an *arabica* strain can be grafted on to a *robusta* root stock. Now in all these cases there have been tremendous advances in recent years in terms of seed technology. Micro and mist propagation methods further help to accelerate progress in the multiplication of clonally propagated material. We can preserve hybrid vigour to some extent in maize through a mechanism called composite production, which enables farmers to keep their own seeds. In the Ernakulam district of Kerala over 30,000 growers of orchids are practising micropropagation through tissue culture. Also the micro propagation of forest tree species through elite material, is an important contribution. Thus, the seed industry has progressed much during the last thirty years, particularly after World War-II. Modern biotechnology has given further impetus to the growth of the seed industry.

In India, the seed industry prior to Independence was largely confined to some ornamental plants and vegetable crops. It was hardly developed in the case of cereals. Even as early as 1960 the difference in price between seed and grain was only Rs.5/– per quintal, thus rendering the raising of crops for seeds economically unattractive. And when we wanted to propagate the high yielding varieties of wheat, fairly fast, then we had to organise a Seed Village in the Khajanwala Block of Delhi State. Everybody took to seed production in the village. When farmers found they got a good price for seeds of dwarf wheat varieties, they developed their own methods of rapid seed multiplication. By dibbling seeds at wide distances they increased the multiplication ratio to over 1 : 100. Assured and remunerative marketing opportunities stimulate farmers' interest in taking to seed production. Even now, in crops like pulses and oilseeds an efficient method of high quality seed production will be the organization of seed villages.

From the first five year plan onwards in India there has been emphasis on seed production and distribution. We now have nearly 100 million operational holdings. This means about 500 million people are dependent in our country upon agriculture for their livelihood. Farmers usually adopt a mixed farming system. This is why, seed production of crops, like fodder crops becomes very important. In the first two Five Year Plan periods, when the community development programme was in sharp focus, the approach to seed production was a highly decentralised one. Government desired that every community development block should have a seed farm and that there must be a method by which the seeds needed by the farmers in each block are produced at the block level seed farms. Unfortunately, these seed farms were run in the manner of government offices. I recall when Mr. C. Subramaniam was the Minister for agriculture in mid-sixties, he once humorously said, "... whenever I travel in the country side, I find that the farm with the poorest looking crops is invariably a block seed farm". Unfortunately poor implementation led to the demise of a good idea, namely decentralised seed production at the village and block levels.

With the advent of hybrid maize, the Government of India set up in 1963 a National Seeds Corporation with technical help from the Rockefeller Foundation. The National Seeds Corporation was initially set up mainly for producing and distributing seeds of hybrid maize, hybrid sorghum and hybrid pearl millet. Several steps were taken by the Government at that time to promote a healthy seed industry. A system of coordinated yield trials was developed by the Indian Council of Agricultural Research for assessing the yield potential of new strains. A Central Variety Release Committee was established as a part of the Seed Act. Its job was to regulate in a disciplined way the release of better varieties of crops. Steps were also taken to promote a seed production and distribution chain at three levels. The breeder's seed, which is the purest form of seed provided by a breeder or a breeding institute, is multiplied and made into foundation seed. Then it is given to registered seed growers whose field can be inspected and the seeds certified. Thus there is a hierarchical order at 3 levels, breeder's seed, foundation seed and certified seed, and finally the certified seed is the one which is sold to farmers. The National Seeds Corporation has unfortunately not been able to operate in a commercially viable manner. I noted from NSC's reports that it had run into a loss of Rs. 88 million by 1988. The State Seed Farms Corporations had run into a loss of Rs. 95 million. Fourteen different State Seed Corporations exist now. Out of them, eight are showing profit and others are showing losses. About 38 per cent of the certified seed in the country comes from the public sector. 50.54 lakh quintals of seeds were certified in 1988. The production may be about 25 lakh quintals of certified seed now. When we produce 180 million tonnes of food grains alone, the seed requirement will come to about seven to ten per cent of the total quantity of food grains produced. In other words, the predominant seed industry in India is the seed produced by farmers' themselves and which they keep for their use. Farmers, when trained, can produce good seeds for themselves, even in vegetatively propagated crops. In potato, for example the seed crop can be raised during the aphid free season. Thus, the major seed industry in our country will continue to be the farmers' seed industry. The private sector seed industry is also growing rapidly, but they largely deal with high yielding varieties or hybrids or ornamental plants or horticulture crops. We should foster the growth of a highly decentralized seed industry, since this will help to provide high quality seeds or planting material of location-specific strains. Location-specific varieties provide the foundation for sustainable agriculture. Seed technology also provides opportunities for women and men belonging to landless labour families to achieve value addition in their work.

I would like to deal briefly with the international aspects of the seed industry. Success in plant breeding depends on access to a wide range of genetic variability. As I mentioned earlier, we owe to rural women and men the large variability existing within a species, particularly in plants of economic value. They have preserved them with care and concern. While plant breeding in developing countries has remained largely in the public sector (Universities and State-financed Research Institutions), such work in industrialised countries is largely in the hands of private companies. Private companies obviously have to make profit for their survival and growth. In order to reward private companies for their efforts, different systems of intellectual property rights (IPR) are in vogue in different countries. In Europe and North America, this has increasingly taken the form of "Breeders Rights" regulated under a convention evolved by the Union for the Protection of New Varieties of Plants (UPOV). Some form of reward and recognition for private plant breeders and plant breeding companies has been in vogue in several countries in Europe like the Netherlands for over 50 years. The UPOV convention maintained until 1991 the fundamental principle of unrestricted access to genetic resources. It had provisions for Breeder's Exemption (i.e., breeders can use a protected variety for creating and comercializing a new variety) and *Farmer's Privilege* (i.e., farmers are permitted to multiply propagation material of a protected variety to be used for growing crops). However, the revised UPOV convention of April, 1991 has eliminated these two provisions. The new UPOV convention stipulates the extension of the industrial patent system to plant genetic resources. The trend towards patenting of plant varieties is because of the perceived inadequancies in PBR legislation in some countries like the USA. UPOV and WIPO (World Intellectual Property Organisation) have concluded that the two systems are complementary and that patents are highly appropriate for biotechnological inventions. These can then be incorporated into plant varieties protected by PBR legislation. Such trends in the the industrialised nations led developing nations to propose the concept of *Farmer's Right* in the forum of FAO. Farmer's Rights are defined as the rights arising from the past, present and future contributions of farmers to conserving, improving and making available plant genetic resources, particularly those in the centres of origin of diversity. This concept was approved by all Member Nations of FAO in 1989.

The Keystone International Dialogue on Plant Genetic Resources which I chaired between 1988 and 1991 has offered suggestions for developing a recognition and reward system for informal innovation represented by the concept of Farmer's Rights. The Global Convention on Biological Diversity stipulates that "States have sovereign rights over their own biological resources". Thus, we are confronted today with the need for determining the relative contributions, in terms of social recognition and economic reward, of those who have contributed genetic resources and those who have recombined such resources in a manner which gives rise to commercially successful varieties and hybrids. As the role of the private sector in the seed industry is increasing in India, we should not lose further time in addressing this issue. This is important not only for the growth of the Indian Seed Industry but also for regulating international relations on mutually beneficial terms.

The issue of patenting life forms has aroused much debate in recent years on ethical and equity considerations. The clauses on Intellectual Property Rights in the Dunkel draft text of the General Agreement on Trade and Tariffs (GATT) have evoked mixed reactions. Article 27 of the Dunkel draft makes it necessary that some form of IPR system is introduced in plants. Patents themselves are not obligatory but either patents or an "effective *sui generis* system" such as Plant Breeders Rights are compulsory. The U.S. objection to the Global Biodiversity convention in 1992 related largely to the provision of effective and adequate intellectual property rights. (President Clinton has since signed the convention).

The effective *sui generis* system referred to in the Dunkel draft has in mind the plant breeders rights (PBR) enshrined in the UPOV convention. One of the requisites for the successful conclusion of the Uruguay Round of GATT is an agreement on this issue. The 1991 amendments to the UPOV convention have strengthened breeder's rights and ignored farmers' rights. It is in the light of this situation we have to consider the nature of protection essential, first to protect the IPR of Indian breeders, second, to stimulate private investment in the seed industry and third, to ensure that farmers's rights are protected. A committee set up by ICAR under the Chairmanship of Dr. S.Z. Qasim has given a set of recommendations. These can form the starting point for formulating our national policy on this vital issue. Inaction will lead to our losing the full benefits of the excellent plant breeding capability of the public and private sectors of our country on the one hand, and to denying the emergence of a recognition and reward system for the informal innovation systems of farm women and men. Also, we will not be able to operationalise our sovereign rights over our genetic wealth.

The spread of patenting of biotechnological inventions and more particularly of naturally occurring genes since the mid-eighties will hamper free germplasm exchange. The biotechnology industry of developed countries is keen to strengthen patenting laws at all levels-product, process and product by process. Formerly, only finished varieties (genotypes) were registered for benefit under breeder's rights. Now there is pressure for patenting individual genes and DNA sequences. The US Supreme Court in a ruling stated: "The laws of nature, physical phenomena, and abstract ideas have been held not patentable. Thus a new mineral discovered in the earth or a new plant found in the world is not patentable suject matter". Taking this analogy, it can be argued that naturally existing genes are also not patentable. The role of lawyers in the breeding industry is likely to increase and it will not be surprising if soon there are more lawyers than geneticists associated with the breeding and genetic engineering enterprise. The casualty of this trend will be global food security.

Progress in the world of plant breeding and seed development until now has largely been due to unrestricted exchange of genetic stocks among breeders. The success of the new wheat and rice varieties is to a considerable extent due to the incorporation of the Norin dwarfing genes in wheat from Japan and the Dee-Gee-Woo-gen dwarfing gene in rice from China. Unfortunately, the era of cooperation in developing varieties which can strengthen global and national food security systems seems to be nearing its end. The poor countries and the poor in all countries will suffer most from a decleration of progress in plant breeding.

Seeds provide the most effective safety net for resource poor farmers. Good seeds, combining tolerance or resistance to a wide range of biotic and abiotic stresses, can alone help small farmers to achieve the twin goals of increasing productivity and decreasing the cost of production. Our seed policy therefore should contain the following components.

1. Strengthen ongoing efforts in *in situ* and *ex situ* conservation of biological diversity and revive the role of farm families in the *in situ* conservation of land races and earlier cultivars.
2. Strengthen plant breeding programmes by integrating genetic engineering techniques with conventional breeding procedures and set up for this purpose Genetic Enhancement Centres, which can provide to grass root level plant breeders breeding material containing novel genetic combinations of value in sustainable agriculture.
3. Enlarge training in seed technology and enable farm families particularly women to grow and retain good quality seeds or planting material to the maximum extent possible. Develop decentralised seed production techniques including micro-propagation and support them with key centralised services.
4. Promote the establishment of seed orchards and nurseries for the propagation of vegetatively propagated plants and train women and youth, belonging to landless labour families in budding, grafting, hybrid seed production and other techniques which will help to add value to the time they spend in work. Also, set up micro-propagation facilities, where needed, in seed villages for multiplying planting material of flowers and fruit trees.
5. Develop a *sui generis* system of varietal protection which will help to prevent a situation where our seeds are freely available to others but we have to pay royalty to seeds obtained from outside and which, for the first time in the world, will provide for recognising and rewarding farmers' rights.
6. Refuse to be a party to any international agreement which (a) will deny farmers the right to keep and plant their own seeds (b) does not recognise the role of informal innovation and (c) impairs free exchange of germplasm among scientists engaged in strengthening global food security.
7. Develop methods of protecting and utilising in an ecologically sustainable manner our sovereign rights over genetic wealth.

Because of our agro-ecological diversity and plant breeding capability and because of the practical skills of rural women, we have the potential for fostering the growth of a dynamic seed industry, both for meeting domestic needs and for exports. Our aim should be the provision of good quality seeds in all crops to the 100 million farming families in our country and to capture about 25 % of the world trade in seeds and planting material during the next 15 years.

This will help to provide numerous skilled jobs in villages and foster on employment triggered economic growth strategy. The issue of unrestricted exchange of germplasm in crops is of importance in South-South collaboration, even more than North-South relationships, since most of the mega-diversity areas occur in the South. Hence, matters relating to the seed industry need careful consideration at G-15 meetings.

As I mentioned earlier, seed is a powerful safety net for resource poor farmers. The seed industry can become a powerful engine for human resource led economic growth. Rural jobs involving value addition to time are the need of the hour in countries like ours with a predominantly young population.

In any policy on job-led economic growth in rural areas, seed technology including the application of biotechnology will have to occupy a pivotal position. Public policy with reference to national action and international cooperation will have to be based on this realization. There should be agreement among countries in the South as well as in North-South dialogue on methods of continuing the unrestricted exchange of genetic stocks related to food and nutrition security and to adaptation to potential changes in temperature, precipitation, sea-levels and ultraviolet-β radiation. Methods of harmonizing the imperatives of public good and private profit will have to be developed. Research at the International Food Policy Research Institute has shown that the number of malnourished children in developing nations increased from 166 million to 188 million between 1975 and 1990. Scientists will be remiss in their social responsibility if they do not insist that meeting basic human needs should be kept above purely profit motives. At the same time, there is need at the political level for a total commitment to decentralised seed production and to farmers raising their own seeds in a scientific manner.

At the international level, any *sui-generis* system developed for rewarding formal innovation should have built-in provision for recognising and rewarding informal innovation. The chain of progress in plant breeding will get broken if the last link in the chain alone receives attention. It would be in the national interest if the Goverment of India sets up a Committee with representatives of the major stakeholders in the field of conservation, evaluation and utilisation of plant and animal genetic resources to prepare within 3 months a draft national policy paper.

Patenting Living Organisms and Their Parts: The Point of View of Science Administration

R. Grunwald

A. Introduction

Patenting of biotechnological inventions is no longer an exotic affair but every-day-routine in the patent offices around the world. The market for the respective products is already now at an estimated volume of about 10 billion Dollars per year with high growth potential. One of the prospective main areas of exploitation – medical use – is still in its initial phase.
Molecular biology has changed the character of life sciences in general and especially biology from science formerly based mainly on experience and description to a natural science much more comparable to physics and chemistry. It has made its subject part of the exact sciences. Besides, because of the speed of its development, it has in many areas changed from a subject of research to an instrument in a very short time. What a couple of years ago was frontline research has since become knowledge and method readily available for broad use. A new kind of science bred a new kind of scientist. Although contemplation and patient observation still rank highly, competitive setting of research aims and reaching them before the runner-up has gained importance. The division of labour and form of cooperation within the life science community had to be developed and are still on the move.

What for the experts was a logical development took the public by surprise: When the European Patent Office (EPO) published its decision to patent the "Onco mouse/*Harvard*" on May 13, 1992[1], public discussions only started. This is true, too, for patenting of plants and parts of living organisms, especially of the human genome. The phenomenon of public awareness catching up with expert knowledge and opinion is not new: The changes of our technical world have created many examples, atomic energy and the role of chemical products for the environment are only two recent ones.

In the following I shall outline the consequences of the patent issue from the point of view of the *science administrator*: What is already established state of the art and has to be dealt with like facts, and what is still necessary. Planning, deciding, controlling: The patent issue is of importance to all these levels of research management. Every craftsmen has to muster his instruments before he sets to

---

[1] European Patent Application No. 85 30 4409.7, rejected by an Examining Division of the EPO, decision of July 14, 1989, OJ EPO 1989, 451; remitted by the Technical Board of Appeal to the Examining Division, T 19/90, OJ EPO 1990, 476.

work: Access to, working with and exploitation of biological knowledge and especially inventions has to be considered by research administration.

For a research administrator – not only of a life science centre – the main task is to *secure the resources necessary to operate the unit* he serves and to *create an environment in which science can flourish* and *scientists can work to the best of their potential for the mission of the research unit.* The framework in which he works is formed by a highly complicated set of factual, social and legal facts and rules. Highly complicated partly because of the nature of science, exploring new borderlines of knowledge and human options, partly because of the many systems he has to interact with – not only within his science community but with the social, political and economical world outside, too. One of his tasks is to function as an early warning system to identify necessities to develop for example new rules or change existing ones because of developments he learns from the scientists he cooperates with. He has to act partly, too, as an interpretor between the science community and the social, economical, political and legal environment.

The issues raised by patenting living organisms and parts of these are from this point of view

- *concerning the interaction within the science community:*
  free excange of knowledge and materials vs. protection and exploitation,
- *concerning organisation:*
  freedom of action vs. organisational precautions and defined steps for instance before publication,
- *concerning types of science:*
  project/result orientation vs. search for truth.

From the point of view of science the priority has to be: *As much freedom of action for science as possible, as much protection of other interests as necessary.*

B. The State of the Art: Where Do We Stand?

I. Overview

The existing decisions and expert opinion can be summed up as follows.

1. *Patent law does not exclude patents on plants or animals*, only restricts such patents to levels higher than the lowest taxonomic levels.[2]
2. *Life reduced to its chemistry is patentable* as long as the prerequisits of Art. 52 European Patent Convention (EPC) are fulfilled: *Inventions* which are *suscectible of industrial application*, which are *new* and which *involve an inventive step* – and are not excluded by the relevant rules.[3]

---

[2] See "Rote Taube" BGH GRUR 69, 672 T 19/90 EPO 1990, 476 Onco-mouse/Harvard Kinkeldey, Ursula, Die Patentierung von Tieren, GRUR 93, 394; Vincenzo Di Cerbo, Die Patentierung von Tieren, GRUR 93, 399.

[3] Jaenicken, Hans-Rainer, Die Patentierung von Biotechnologie – Erfindungen beim EPO with references, GRUR 92, 327

3. Merely finding a *substance occuring in nature* is a *discovery* and regarded not as invention, Art. 52 II (a) European Patent Convention (EPC). The distinction towards patentable substances is drawn along the line: "Isolation of the substance occuring in nature from its surroundings by a particular process developed for obtaining it." And: Not only the process itself, which is clearly man-made, is patentable – the substance isolated, too.
4. All *restrictions of patentability are exceptions from the rule* and therefore to be construed narrowly.

In the following I shall first discuss *whether the state of the art is a clear – cut solution* of defined problems or whether open questions remain, then I shall trie and point out which *consequences – possibly problems – develop out of this situation.* Finally, I shall discuss *possible lines of action* to minimize possible problems and keep open the free flow of information and ideas.

II. A System On The Move: The "State Of The Art"

The legal system is like a living organism: It has to adapt to changing circumstances, find solutions for new questions and exchange old answers for new ones if necessary. Besides, some basic truths and rules have to be respected like: "Do not harm anybody unless just cause requires you to do so."

In analysing the question whether there are already clear – cut solutions in the field of patenting living organisms and their parts we have to look at the relevant *patent law, decisions of the patent authorities and courts,* finally find out whether *alternatives* remain to be discussed.

1. The Theory Of Isolation: End Of The Discussion?

The distinction between patentability of inventions and non-patentibility of substances found in nature is drawn along the line of "*isolation*": As clear as this distinction is for the *procedure*, the process of isolation, as long as it meets the standards of patentability set forth for example in Art. 52 EPC, as doubtful it remains for the *isolated substance*, the result of the procedure itself. If today the microscope was invented, there is no doubt about the patentability of the procedure, the instrument, but would the exploitation of the life found in the famous drop of water now belong for the period of patent protection to the inventor of the microscope? Making something existing in nature visible, tangible, smellable, audible, tastable, in short accessible for the human senses and understanding, does not necessarily mean he can "own" it, exclude others from its use even for a limited time: Not only the process, but the substance found, too, should meet all the standards set forth for the patentability of inventions.

This up to now seems not to have been distinguished clearly enough. If the procedure was found to be patentable the result was more or less automatically deemed to be patentable, too. For parts of the chemistry of life this is questionable. Firstly, whether you discuss the full genome of an organism or only part of it after isolation, it is there, you only make it *accessible to human perception.* Even if you open it for manipulation it is still already there and you only open a natural substance to human manipulation. This might not have been as clear some years ago when the *pioneering findings of biotechnology* where not clearly recognizable as merely the first links of a long chain of discoveries which in a short time led to quickly deciphering the chemistry and biology of life. *Now, the situation has changed.* We know now that access to the inner structure of genetics is possible and that all findings are only steps on the way to a more and more complex knowledge and understanding of life and its processes. This means, too, that the question of protecting even single steps has possibly great importance. Let us argumendi causa presume there was something like the "*life switch*", the key to prolonged or even eternal life. This could be true for the opposite, too: Who deciphers for example the apoptosis – mechanism, the self – destruction of a cell, might have a potent agend at his disposal. Not only single steps, but ownership or economic command over all of or important parts of the human chemical and biological heritage, too, need a closer look. The mere possibility to formulate these questions illuminates that *parts of the human body* – tiny and untangible as they might be – *are something at least very special if* not so unique that they need especially careful consideration under all relevant aspects. Economic exploitation is only one of those aspects.

To sum up: The doctrine of "isolation" as a borderline between substances found in nature and patentable inventions is valid as far as the *isolating procedure* is under scrutiny. The result of the isolating process has to meet all the prerequisites necessary for a patentable finding, too – and in many cases it will be a mere *finding of substances occuring in nature.* Only if by the process of isolation or later the substances are *manipulated,* used by man in a way that changes the relevant substances or finds new functions and thus made "new" they are *not only "man-found" but "man-made"*, thus possibly patentable[4]. The use of restriction enzymes would have been a patentable invention, the pieces cut from a longer DNA – chain as such are under this distinction merely naturally occuring substances which are made accessible through a patentable procedure unless a new, surprising use is found – "discovering" nature's functions is not enough. The newly invented flint-stone knife would have been patentable if man had already used patent law – the pieces of meat cut from the mammoth would have been naturally occuring substances and not patentable unless they were for instance usable as medicine.

---

[4] The Swiss Bundesrat exempts unchanged organisms, for example genes. Decuring in Nature from patentability, see Biologie und Patentrecht, Bericht, Eidg. Justiz- und Polizeidepartement, Bern 1993, p. 42.

2. Where There Is A Will, There Is A Way? Tests To Be Met Before A Patent Is Granted

From the point of view of science administration some other issues have to be discussed under patent law besides the problem of the isolating manipulation of substances occuring in nature.

a) Under Art. 52 (4) EPC *methods for the treatment of the human or animal body by surgery or therapy and diagnostic methods practised on the human or animal body* shall not be regarded as inventions which are susceptible of industrial applications. This provision shall not apply to products, in particular substances or compositions, for use in any of these methods. In our context – use of biological procedures and substances – the distinction between "use on the human body" from other uses will be of importance. Treatment for example of body components (tissues, fluids) could be patentable as far as these components could not be returned to the same body. For the research unit, even more important: for a hospital, it could be necessary to distinguish for example between plasma to be returned after treatment to the same body and plasma to be stored in a plasma bank. Could a pharmazeutical firm bind a hospital not to use its procedure to be used for example on plasma which will be returned to the same body? This does not seem a very practical distinction – you would have to make sure by organisational means that plasma stored in a plasma bank could not be returned to the donor even though under medical aspects this could be necessary. The less we can distinguish between procedure and product – as will be the case with gene-diagnostics and gene-therapy – the more difficult – and doubtful – this distinction could become.

b) Under Art. 53 (b) EPC *plant varieties* are excluded from patentability. This is in accordance with Art. 2 "International Convention for the Protection of New Varieties of Plants" (UPOV) and makes clear that there is at the moment no possibility of *double protection*: Either there is protection under patent law or under national plant variety protection acts. It should, however, be noted that an amendmend of Art. 2 UPOV was suggestetd in Geneva in early 1991 to give up the verdict of double protection, but this amendment is not yet accepted. So the exclusion of double protection is still the law[5].
Under this regime it is argued that patent protection could cover in the field of botany

   - anything not defined as plant variety in the sense of the UPOV (suggesting broader claims),

[5] See T 49/83 "Propagating Material/CIBA GEIGY", OJ EPO 1984, 112 sequ.: "The legislator did not wish to afford patent protection under the European Patent Convention to plant varieties of this kind (i.e. all cultivated varieties, clones, lines, strains and hybrids which can be grown in such a way that they are clearly distinguishable from other varieties, sufficiently homogeneous, and stable in their essential characteristics), whether in the form of propagating material or of the plant itself". In accordance with T 320/87, "Hybrid plants/Lubkizol", OJ EPO 1990, 71 ff.

- parts of plants,
- tissue cultures, and
- plant cell lines.[6]

This again raises the question of substances occuring in nature: From the point of view of science administration all these examples as such lack the characteristics of an invention, as long as there is no "human intervention" which is an *inventive step as far as the substance itself* is concerned: The biotechnological process is clearly "man-made", but not the substance isolated or found, unless it meets the standards of an invention as a substance: To be susceptible of industrial application, be new, represent an inventive step and not be excluded from patentability. Besides, adequate protection at least for plant varieties and their parts could be discussed under a development of the plant varieties protection system: This covers varieties not "invented" but "discovered", too, and is especially tailored for propagation and not primarily technical multiplication. As the use of propagating material of a protected plant variety for breeding a new variety is free and does not need the consent of the owner of the protected plant variety[7], and as under plant varieties protection law as well as under patent law private use for non-commercial purposes is free, the only addendum needed would be a *research clause* stating that experimental use is free, too.

c) Patents are excluded for *animal varieties*, Art. 53 (b) EPO. The wording and the meaning of this clause are under dispute. The German version "Tierarten" should – and this is in conformity with plant variety protection law – be read as "Tierrassen", i.e. draw the line between protection and no protection at the lowest taxonomic level. The deeper reaching question is: "What does this exclusion mean?" This was discussed in the various decisions concerning the "Onco-mouse/HARVARD." The Examining Division, OJ EPO 1989, 451 ff., held that "Art. 53 (b) EPC excludes patent protection for animals per se in general, and not only if a particular variety is claimed." This decision was not upheld by the Technical Board of Appeal[8]. It argued that the general rule of Art. 52 (1) EPC – European patents "shall be granted" to inventions – implied that exceptions have to be construed narrowly. It then states that the preparatory documents are completely silent as to the purpose of excluding "animal varieties" – in contrast to "plant varieties". It then considered the fact that no other industrial property right was available for the "inventor" and that an inventor in the field should be able to find reasonable protection for his "invention". It finally pointed out that Art. 53 (a) EPC had to be considered and stated: "The genetic manipulation of mammalian animals is undeniably problematical in various respects." It then remitted the case back to the Examining Division for further examination which decided to grant a patent on

[6] See T 320/87 "Hybrid Plants/LUBRIZOL", OJ EPO 1990, 71, and T 49/83 "Propagating Material/CIBA GEIGY AG BASEL", OJ EPO 1984, 112.

[7] § 10 last sentence German plant variety protection act.

[8] T 19/90, OJ EPO 1990, 476, "Onco-mouse/HARVARD".

four claims relating to "transgenic non-human mammalian animals"[9]. It enclosed reasons for allowing the case with its Rule 51 (4) EPC Communication dated October 4, 1991[10]. After stating that the subject matter of the claims was not covered by Art. 53 (b) EPC and therefore possibly patentable it discussed in depth the question whether Art. 53 (a) EPC barred the claims under the aspects of public order and morality. It weighed the possible risks for the environment and the possible suffering of the animals against the interest of mankind in the invention's usefulness. It decided that the benefit to mankind outweighed the negative aspects. It finally stressed that this decision covers *solely the present case*. Presently the opposition proceedings before the EPO are under way: Opposition has been raised by 16 parties ranging from the British Union for the Abolition of Vivisection to the Zentralverband der Tierschutzvereine Österreichs.

The Technical Board of Appeal is correct in stating "that for aminals – unlike plant varieties – no other industrial property right is available." But: It is correct, too, to state: "For the time being". Up to now there was *no necessity to protect cattle or pets by industrial property law*, as there was hardly any "industrial interest" involved. The only protection needed was for agricultural interests and those of pet breeders. This is about to change. Especially animal models for human illnesses will be developed in large numbers as transgenic animals and with high economical potential, thus starting a new line of experimentation and, finally, industrial use. To see this development under the aspect of – even bio – technology would not sufficiently take into account the subject "living animal" which is not for instance mechanically multiplied but breeds. From the point of view of research administration the hint of the Technical Board of Appeal should be carefully considered: The *development of an industrial property right especially for animals following principles developed for plant varieties protection* – free use for the development of new races/varieties for example – could be a well-suited solution for the problems at hand[11]. This could possibly be incorporated into the patent acts and conventions or, like plant varieties protection, be laid down in separate acts and conventions.

d) *Essentially biological processes for the production of plants and animals are* excluded from patentability under Art. 53 (b) EPC. The argument of the Technical Board of Appeals[12], patents are grantable for "animals produced by a microbiological process" is easily misinterpreted to mean "involvement of a microbiological process secures patentability." Clearly the individual animal which was genetically manipulated is the "product" of human manipulation, thus "man-made". But already the next generation is bred and the argument

---

[9] EP-B1 0 169 672.

[10] OJ EPO 1992, 588.

[11] Hesse, Hans-Gerd, Der Schutz der züchterischen Leistung + die GRechte, GRUR 71, 101

[12] T 19/90, "Onco-mouse/HARVARD".

that this is essentially a biological process should not be put off too easily. The decisions concerning the parallel problem – exclusion of essentially biological processes for the production of plants – show that the test is the extent to which there is technical intervention by man in the process. This, however, means, that the process of breeding is primarily a normal biological process and not easily patentable as such. The patent would have to cover the "subject matter", i.e. the animals – and we discussed above that a special form of industrial property protection for living animals seems advisable.

e) The EPO follows the absolute novelty standard: To be patentable an invention has to be new and go beyond the state of the art, i.e. "everything made available to the public by means of a written or oral description, by use, or in any other way, before the date of filing of the European patent application", Art. 54 (2) EPC. From the point of view of science administration prior publication is the most important – and most common – stumbling stone for a successful patent application. The U.S. American patent law for example is much more lenient with an inventor: It accepts a patent application as new up to one year after the first publication of the invention. This should be made compatible all over the world as soon as possible – the different standards are not up to date with the fast dissemination of informations[13]. The one-year-"grace-period" after publication for deciding whether a patent application should be filed is much more adequate for the actual process of innovation and decision making in this field: The scientist-inventor – and the research centre as the owner of the possible invention under for instance German patent law – can find industry partners interested in the new findings. It takes time to decide whether an invention is a sound basis for the investments needed not only to file the patent application but, much more important, to run the risk of a development.

f) The necessary *inventive step* is defined by Art. 56 EPC: If, with respect to the state of the art, the invention is not obvious to a person skilled in the art. To test the *non-obviousness*, first the technical problem has to be defined for which a solution is claimed to be found, then it has to be discussed whether the suggested technical solution was obvious to a skilled person. Biotechnology is an especially fast developping art. When the pioneering steps were taken they were revolutionary and easily accepted as new, even describing their findings by procedure. The more the new technology was established in laboratories and industry around the globe, the higher rose the standards of the state of the art, the less convincing becomes the theory of the isolating human intervention. We have to accept that humans moved into the microcosmos and manipulate it. When Columbus landed on the beaches of the "New World" – for him a great discovery, for the Indian who met the first Spaniard it was "a bad discovery", as the German philosopher Lichtenberg remarked already 200 years ago – he did not only bring the cross but also the Spanish flag to take

[13] See also Biologie und Patentrecht, Bericht, Eidgen. Justiz Polizeidepartement, Bern 1993, p. 9.

possession of the land for the Spanish Crown. When the first American set his foot on the moon, he, too, brought a flag – but not to take possession of the moon for the United States of America but to symbolize a new step for mankind, proudly taken by an American. So let us accept that we moved into the microcosmos, too, but let us find mechanisms to deal with these explorations and discoveries and not imitate Columbus and the conquistadores who followed him and tried to own – and actually plundered – what they discovered.

g) Art. 57 EPC gives a broad definition of the *utility standard* which opens patent protection for any commercial production or use including agriculture. Only in cases where there is no definable utility at all or only hope to find such a utility one day no patent will be granted. This standard is not met for instance as far as the NIH-claim to patent certain unspecified cDNA sequences is concerned.

3. Last Ditch Defense: Ordre Public And Morality

No European patents are granted for inventions contrary to ordre public or morality, Art. 53 (a) EPC. The fact that decisions have to discuss these general clauses as the basis for a decision indicates that very basic questions are involved: Normally either one decision which then is the legal state of the art sets clear rules or – if not only very few decisions have to deal with these questions – legislation is not far away. General clauses are the legislative instrument to allow movement where too detailed rules would otherwise stifle developments, on the other hand they protect basic principles of a legal system against too quick changes and easy circumvention. In German unfair competition law the general clause making illegal all competitive acts contrary to morality (§ 1 UWG) is such a defense against all acts against the spirit of fair competition although not specifically prohibited by special rules. As imagination and fantasy allow to find ever new forms of competitive acts legislation will not possibly succeed in regulation every detail of possible violations of fair competition. On the other hand, if very basic principles are involved, the legislator will not wait indefinitely watching litigation grow and with it insecurity and unrest among those concerned or the public, but act by setting specific rules.

These considerations can be applied to the general clause Art. 53 (a) EPC. Contrary to *ordre public,* i.e. *basic principles of legal order*, will foreseeably be only few patent claims: Safety seems to be the most relevant basic principle to be considered in this context. As this area is well covered by legislation, patent decisions can well be oriented along the guiding lines and basic rules of this legislation. More difficult is the question of morality. As an exception to the rule of Art. 52 EPC the restriction of Art. 53 (a) EPC has to be interpreted narrowly: The mere risk of a new technology for instance can not simply outweigh the right of the inventor to become the owner of a patent. The interests involved have, however, to be balanced carefully. This was done in general terms by the Technical Board of Appeal in the "Onco-mouse/HARVARD" case, T 19/90, OJ EPO 1990, 476, and discussed in detail by the Examining Division, see OJ EPO 1992, 588.

Three different interests are distinguished: A basic interest of mankind to remedy widespread and dangerous diseases, on the other hand the environment, which has to be protected against the uncontrolled dissemination of unwanted genes and finally the animals, cruelty against which has to be avoided. The Examining Division summed up its reasoning: "In the overall balance the Examining Division concludes that the present invention cannot be considered immoral or contrary to public order. The provision of a type of test animal useful in cancer research and giving rise to a reduction in the amount of testing on animals together with a low risk connected with the handling of the animals by qualified staff can generally be regarded as beneficial to mankind. A patent should therefore not be denied for the present invention on the grounds of Art. 53 (a) EPC. It should be stressed that the above considerations apply solely to the present case and that other cases of transgenic animals are conceivable for which a different conclusion might be reached in Art. 53 (a) EPC."

The patent decisions concerning animals were up to now only a few cases worldwide. Four years after the "Onco-mouse" patent was granted in the U.S.A. in 1988, four other patents were granted for mammalians (December, 1992). Today, applications for more than 80 other "animal-inventions" are filed in the U.S.A. alone – a dam is about to break. From the point of view of research administration, especially in health research, free and fast exchange of information is of utmost importance. We now face a situation which for the years to come is characterized by insecurity and impending litigation. Not only will the "Onco-mouse" case be followed through in the procedure before the EPO, national litigation will be the next step. Considering the parties involved, it is not difficult to predict that for many "animal-patents" granted such litigation will follow – possibly with different outcome in different national courts. This is a situation which cannot be calmly accepted by the bio-medical science community. It has to discuss the issues involved on an international level and formulate its recommendations how to cope with these issues.

4. Summing up the Legal Situation

To sum up the discussion of the legal situation: The state of the legal art is under scrutiny; the problems to be solved are new under the aspect of quality ("mammalian-patents"; parts of the human identity as patent subject; "man-made" animal models for human illnesses) as well as under quantitative aspects (first "mammalian-patent" granted in 1988, four others in late 1992, more than 80 U.S. patent applications pending).

C. Issues to be Discussed

Patent law is an important instrument of the innovative process: The necessity for the inventor to make his *deal with the public and publish his invention* is aimed at the dissemination of information and advancing technical development. Without patent protection part of the newest technical and scientific achievements would possibly be withheld from the public to secure later exploitation or to protect the lesser "state of the technical art". This is true, too, for bio-medical inventions.

Discussion is needed in respect to complications, even dangers for scientific research caused by the efforts to patent results in this area.

1. The Interaction Within The Science Community: Free Exchange of Knowledge vs. Protection And Exploitation

Scientific research funded by public means is aimed at the free exchange of knowledge, materials and results: The division of labour in our research system hands the – purpose-free – search for knowledge over to the public part of the system. Market- or even product-oriented research on its way to development is taken care of by the commercial part of the system. The interface between these parts of our research system is affected and the borderline changed by the growing tendency to trie and protect results of bio-medical research by patenting inventions in this field. Project-orientation and short-term goals gain more importance. Free exchange of informations, techniques, materials and results is no longer possible without not only risking as formerly the priority of publication but serious economic loss: Safeguards have to be developed to balance the interest of protecting inventions and this free exchange so vital for biomedical research on a global scope.

2. Internal Organisation: Freedom of Action vs. Organisational Precautions

Research organisation in bio-medical research is highly decentralized. There are only few large efforts which function "top down" like the U.S. human genome project. Normally a rather small unit in comparison with for instance physics or engineering research works "bottom up" on highly self-defined subjects. The integration into the relevant science community and setting of priorities takes place through co-opting a leading scientist with his topics, or peer-review before publication or, for instance in a bio-medical research centre, through program discussions and – after then setting priorities – allocation of means. Success is measured mainly by publication in peer-reviewed periodicals, invited lectures and working in a prestigious environment, be it university, be it research centre or industry. Only in industry inventions and patents were considered a main aim of work, in publicly funded research they were more a by-product of research oriented towards long-range and high risk aims not yet close to a market or even product. With the growing tendency to trie and patent results of bio-medical research the scenario and the actors are about to change. Protection of results becomes important, materials and informations are only given under secrecy and non-proliferation clauses as well as only non-commercial use – clauses which characterized the cooperation with industry are now introduced into the cooperation between universities and publicly funded research centres. Now, both the parties interested in commercial exploitation – science administrators and government officials responsible for the money spent on science on the one hand, and industry on the other hand – urge scientists to protect the intellectual property which represents the result of the expenditure of public means and individual scientific merit. Much more centralized organisational features would have to be installed: Rules for publication have to be introduced to prevent the loss of novelty through publication without prior patent application; secrecy has to be secured in the individual laboratory working with substances under binding agreements, materials have to be kept under lock and bookkeeping is necessary to be able to account for the substances. Reports begin to play a much bigger role than before.

3. Result Orientation vs. The Search For Truth: A Change For The Better?

The growing tendency to look for patentable inventions in bio-medical research is a temptation for research administrators as well as for some scientists: Management seems to be furnished with additional success – parameters and possible income, scientists who participate in the revenews smell the sweet smell of money. The problem, however, remains that the outcome of the efforts to generate patentable inventions is highly uncertain, certain on the other hand is the investment needed in the first place and the changes in working conditions and orientation of scientists it brings about. Basic research needs a long breath and patience, orientation towards patentable inventions lacks this quality as it is under pressure to minimize the risk-cost relation. This does not mean that result-orientation is bad: It is, however, necessary to differentiate between the search for truth and short-time projects. These can not substitute the former, attractive as it may be to be able to patent results. Patentable inventions are not seldom the unexpected fruits of basic research, too, but primarily as welcome by-products: A whole international science community should not without careful considerations give away its heritage of free flow of ideas, informations and materials only because a few more scientists inventors will become rich quickly.

D. Conclusion

From the point of view of science administration it is pleaded to discuss on an international level, first of all European level, the development of patenting living organisms and their parts. An opportunity is offered to do so by discussing the EC – directive concerning the protection of biotechnological inventions.

Patent protection for parts of living organisms should concentrate on the isolating or substance – changing manipulation, the substance itself in its natural function should be considered as substance occuring in nature, unless an unexpected new use of it is found.

An industrial property law covering genetically manipulated animals could be developed modelled after the plant varieties protection and either integrated into patent law or enacted separately.

Plant varieties protection should grant a reserarch privilege.

It should be discussed to (re-)introduce the novelty protection of published informations for one year ("grace-period") in accordance with U.S. patent law into the EPC and national patent acts.

Patenting of Human Genes and Living Organisms
The Current Practice of the European Patent Office

Ch. Gugerell

Introduction

The European Patent Office is an intergovernmental organisation with now seventeen Contracting States and the underlying purpose of the existence of the Office is that protection of inventions may be obtained in these states by a single procedure for the grant of patents. Thus, the Office is an authority responsible for the examination of patent applications and the grant of presumptively valid patents. There are two legal instances in the Office, the Examining and Opposition Divisions on the one hand and the Boards of Appeals on the other. In this contribution reference is made to Decisions of the Boards which normally have guiding function for the first instance in certain matters of law.

Particularly in the field of biotechnology many further decisions of the Boards of Appeals are to be expected in the not too distant future and therefore the present contribution can only reflect the *current* practice of the Office.

Although the present contribution in its title refers to living organisms, this term shall be understood in the following to be plants and animals but not microorganisms.

Patent applications on plants, animals or somatic gene therapy are no longer rare events, the contrary is true: The Office has received more than 600 applications relating to plants of which about 360 contain claims to plants per se. There are about 220 applications relating to animals and about 190 of them have claims to animals per se. Within less than a year the number of animal cases has doubled, which indicates that there is a dramatic increase of research activity in this field. On somatic gene therapy the Office has already received much more than 50 cases.

Legal Provisions

It is the European Patent Convention (EPC) which is the basis of all legal considerations concerning patentability in the Office. The EPC contains, amongst other things, the requirements for patentability which may be separated in the normal ones (such as novelty, inventive step, industrial applicability or enabling disclosure) and the special ones which are directly applicable to the type of inventions under consideration. Particularly Art. 53 (a) and (b) EPC require close examination in this respect. This article deals with exceptions to patentability and is split into two parts, the first (a) dealing with exceptions on grounds of morality and the second (b) with exceptions in the field of plants and animals. Art. 53 EPC dates back to the Convention on the Unification of certain Points of Substantive Law on Patents for Invention, commonly referred to as the Strasbourg Convention, which was ratified in the year 1963. Obiously this was at a time when the present kind of inventions relating to transgenic plants or animals obtained with genetic engineering methods was inconceivable. This article was taken over unchanged into the EPC and Examiners at the European Patent Office and Members of the Boards of Appeals therefore have to apply and interpret a law with respect to inventions which were not foreseen at the time of enaction by the legislator.

In this contribution Article 53 (b) EPC will be dealt with first. The portion thereof which is interesting in the present context reads as follows:

> "European patents shall not be granted in respect of plant and animal varieties or essentially biological processes for the production of plants or animals; . . ."

It should be noted that the term "variety" is used in this article, there is, however, nowhere in the patent law a definition thereof and resort must be had to the UPOV Convention, which is a law on the protection of plant varieties and which, in its most recent version of 1991 (not yet in force), contains such a definition.

Moreover, it is noticeable that a distinction is made in Art. 53 (b) EPC between plant and animal varieties on the one hand and plants and animals on the other. It can therefore be concluded that it was not the intention of the legislator to exclude from patent protection plants and animals in general but rather only in the form of their varieties. This is in line with the apparent reason for the exclusion of plants. This reason is to be found in Article 2(1) of the UPOV Convention (revised version of 1978 presently in force), which provided for the inadmissability of double protection under the patent and plant variety system for the same subject matter. The reason for the exclusion of animal varieties is not easily apparent from the legislative history.

The portion of Art. 53 (a) EPC which is interesting in the present context reads as follows:

> "European patents shall not be granted in respect of inventions the publication or exploitation of which would be contrary to "ordre public" or morality, . . ."

For determining whether or not a violation of Art. 53 (a) EPC exists the Guidelines for Examination (which is an examining aid examiners have to follow) advise examiners to consider whether it is probable that the public in general would regard the invention as so abhorrent that the grant of patent rights would be inconceivable. As an example for this a letter-bomb is mentioned therein.

Practice Concerning Plant and Animal Cases

Practice under Art. 53 (b) EPC

It was only with the advent of recombinant gene technology and its extension into plant molecular biology that problems of patenting this type of invention arose.

Innovations in this field are normally not limited to specific plant varieties and therefore the protection afforded by the plant variety system was not considered to be adequate. It was patent protection which was sought for these innovations and the European Patent Office had to consider patentability thereof. This led to two early decisions of our Boards of Appeal (T 49/83 published in the Official Journal of the European Patent Office 3/84, 112 and T 320/87 OJ EPO 3/89, 71) in both of which it was held that plants in general are not excluded from patentability and that Art. 53 (b) EPC prohibits only the patenting of plants or their propagating material in the form of a plant variety.

Several years later a similar development occured with animals and the first animal which found its way into patents in Europe was the HARVARD/oncomouse which also gave rise to a first decision of a Board of Appeals (T 19/90 OJ EPO 1990, 476). More or less confirming the decisions on the plant cases referred to above, in the HARVARD/oncomouse case the Board established the legal principle under the EPC that animals in general can be patented, but not "animal varieties". This principle must be seen together with another one also confirmed in this case, namely that in view of the general principle of patentability any exclusion from patentability must be contrued narrowly.

In view of this guidance by the Boards of Appeals Examining Divisions at the European Patent Office adopted a practice which is to allow claims which are directed to taxonomic groups of plants or animals larger than plant or animal varieties *provided* the invention is *APPLICABLE* to such a larger group. It is quite often argued by opponents to animal or plant patents that a plant or animal variety is always represented by individual plants or animals and that therefore a patent claim to "a plant" or "an animal" is also directed to a plant or animal variety and therefore not permissible under Art. 53 (b) EPC. This argument is not convincing because such a claim must be interpreted in the light of the description accompanying the patent claims and if the description clearly teaches a broad applicability of the invention to a large group of plants or animals, it is not permissible to interpret such a claim narrowly as referring to a plant or animal variety. On the other hand, if the description taught a narrow applicability of the invention to a specific variety then a corresponding claim of the above mentioned form would be inadmissible as referring to this variety.

Practice Under Art. 53 (a) EPC

In considering questions of ethics and morality in relation to patents it might be useful to bear in mind some basic points with regard to patents. The following is a quotation from an opposition decision issued on a plant case about a year ago:

> "– A patent does not give a positive right to its proprietor to use the invention but rather only confers the right to exclude others from using the invention for a limited period of time.

– Considering the underlying reasons for the existence of patent law and the nature of patent law in general it appears that patent law is primarily an instrument for promoting research and technical advance associated therewith. It would be an overestimation of the potency of patent law if one considered it an appropriate instrument to suppress research which is dangerous or unacceptable for other reasons to certain groups of mankind. Human curiosity could not be hampered by the fact that patent protection were not available if such research promised an interesting result. If the legislator is of the opinion that certain technical knowledge should be used under limited conditions only it is up to him to enact appropriate legislation outside the framework of patent law."

Following the Guidelines for Examination referred to above the practice of the European Patent Office with regard to Art. 53 (a) EPC was of limited importance. It was only in the above mentioned decision of a Board of Appeals on the HARVARD/oncomouse that the concept of morality in the European Patent procedure has been shifted to a higher priority. There, in connection with the patenting of animals, guidance for considering morality has been given and the Board, in remitting the case to the first instance, ordered to consider certain pros and cons with respect to patentability. In the Board's view patentability under Art. 53 (a) EPC would seem to depend mainly on a careful weighing up of the suffering of animals and possible risks to the environment on the one hand, and the usefulness of the invention to mankind on the other. In this connection it should be noted that the Board concluded that the patenting of animals is not intrinsically unethical but may rather be regarded as morally acceptable provided the balancing exercise leads to a positive result. The first instance, being bound by the Board's decision, then carried out the balancing exercise. The conclusion was that the provision of a test animal in cancer research which may lead to an overall reduction of animal testing – at least this was asserted by the applicant – together with a low risk connected with the handling of the animals was important enough for human society in order to justify patentability. Accordingly, a patent was granted and, after grant, the patent was opposed by seventeen parties which all concentrated on Art. 53 (a) and (b) EPC. The final result of these proceedings – the opposition no doubt will be appealed – will not be available within at least the next four years.

In the more recent past consideration of morality in connection with patents developed a step further insofar as an Opposition Division issued a decision on a plant case where the balancing exercise referred to above with regard to animals was commented on further. The opposed patent related to herbicide resistant plants and the Oponent "Greenpeace" essentially argued that risks in connection with release of the plants into the environment are impossible to predict accurately and that therefore a potential risk to mankind exists. Consequently, in accordance with the HARVARD/oncomouse decision these risks would outweigh all benefits of the invention and a patent was therefore not justified under Art. 53 (a) EPC.

The Opposition Division refused the opposition essentially because the Division did not consider it appropriate to carry out the balancing exercise of the HARVARD/oncomouse decision. The reasons given for this were manifold but primarily the lack of consenus in society on "objective morality" and the inability not only of the opponents but of all experts to prove the extent of the risks and consequently the even bigger problems of a patent office in assessing them should be mentioned. "In conclusion," the decision points out then "the Opposition Division is of the opinion that opposition proceedings before the European Patent Office are not a proper forum for discussing the pros and cons of the genetic engineering of plants in general or the present plants in particular. The present invention does not belong to that extreme category of inventions which could be regarded as so abhorrent to the vast majority of the public as to render the granting of a patent inconceivable and thus excluded unter Art. 53 (a) EPC. Apart from such clearcut cases, the European Patent Office has no mandate to grant or refuse patents on the basis of abstract ethical or moral arguments."

The case has gone to appeal and the question of morality will have to be considered again by a Board of Appeals. The principal question whether the European Patent Office, in applying Art. 53 (a) EPC, should restrict itself to extreme cases only or whether the Office should become a forum for deciding on ethical questions on a whole variety of cases – not only plants and animals are conceivable but also weapons, abortifacients or even herbicides could be the subject of such considerations – is yet to be answered by the Boards.

Practice Concerning Human Genes or DNA Sequences

The question of patenting human genes or DNA sequences only recently has become an issue in the public when it became apparent that the National Institutes of Health (inventor: Craig Venter) in the United States filed a patent application in the U.S.A. and in Europe for more than 2000 human gene fragments without knowing the function of the corresponding genes and consequently without knowing a specific technical use thereof. The Medical Research Center in England followed suit and filed a similar application. It is, however, not this type of patent applications on human genes which is the "normal" one and a distinction must be made between the Venter-type case and the normal ones.

The "Normal" Patent Application on Human Genes

The "normal" cases referred to above are those where human genes or DNA sequences are used as means for expressing in a host certain useful products such as tPA, the interferons, interleukins, growth hormone, insulin or the relaxins. In these cases the respective gene has been identified to encode a certain useful protein and the invention relates to the technical use of this finding. The European Patent Office has developed a practice with regard to this kind of cases which is to allow claims to human genes or DNA sequences because they are an integral part of the overall invention yielding useful products. Obviously the usual criteria for patentability (novelty, inventive step etc.) must be fulfilled in order to be patentable.

The European Patent Office never considered it necessary in the past to apply ethical considerations with regard to claims for human genes or DNA sequences of the above type. The prime reason for this is that patents can only be granted in relation to technical inventions and the effect of a patent on human genes is restricted to the exclusion of others from using the genes in an industrial process. Moreover, the personality of human beings is not affected at all if human genes are protected by a patent for a technical purpose. Quite to the contrary it could be argued that it would be immoral to prohibit patent protection on human genes because there would be less incentive for the provision of valuable medicaments which in large amounts are available only with methods of genetic engineering. Indirectly this practice of not applying considerations on morality has also be confirmed by Boards of Appeals because in none of the cases decided on by Board which related to human genes the question of morality under Art 53(a) EPC has been an issue.

One argument against the patenting of human genes is the alleged possibility that a proprietor of such a patent could inhibit further technical use of these genes in another field, particularly somatic gene therapy, and that in such a case the proprietor would have to be compensated without any contribution. This argument is not convincing because the identification of the gene may be considered a first contribution even for somatic gene therapy. The latter may not be possible without the findings of the first invention. Moreover, dependent patents exist in all technical fields and it has been the political intention of the legislator to provide for these. Otherwise the legislator would have introduced use-restricted rather than absolute product protection and, to a large extent, would have avoided the alleged problem. This has not been done.

Another argument against the patenting of human genes is that the gene is a discovery and therefore not eligible to patent protection in accordance with Art. 52 (2) (a) EPC. In this regard the practice of the European Patent Office is as follows: The finding of a substance freely occurring in nature is a mere discovery and therefore unpatentable. However, if a substance found in nature has first to be isolated from its surroundings and can be properly characterised and is "new" in the absolute sense of having no previously recognised existence, then the substance per se may be patentable provided a technical use has been found for it. Exactly this is the case with regard to the type of inventions referred to above.

All of the above comments on the patenting of human genes refer to the present situation. The question will have to be reconsidered in the near future in an opposition procedure which has been filed by Members of the Green Fraction in the European Parliament against a patent relating to the expression of human relaxin in the host E. coli, which patent also contains claims to the corresponding human DNA sequence encoding relaxin.

The Venter-Type Case

The Venter-type case represents a different category of patent applications and in view of the very early stage of proceedings the European Patent Office has not yet developed a practice at all. These cases relate to human gene fragments with unknown function of the corresponding genes.

As part of the effort to sequence all human DNA under the Human Genome Project Dr. Venter and colleagues of the National Institutes of Health (NIH) began a project during 1991 to sequence partially the approximately 30,000 complementary DNA (cDNA) sequences from the human brain. Venter's group normally sequenced short segments of the available cDNA clones, usually between 150 and 400 base pairs long. These DNA fragments were termed "expressed sequence tags" (ESTs) and the sequence information was stored on a database. The sequencing was done by using conventional automated sequencing techniques.

In June 1991 the first patent application was filed in the U.S. covering 315 ESTs and in February 1992 a second application followed covering a total of about 2400 ESTs. A similar application was filed in Europe. In November 1992 there were already some 20,000 ESTs available in public databases and it is estimated that this will increase by 10,000 a year. No further patent activities on behalf of NIH on these further ESTs are known.

The patent applications are silent on specific uses of the ESTs or the corresponding genes simply because they were unknown at the time of filing. What is disclosed, however, are hypothetical uses for obtaining the corresponding genes, the proteins encoded thereby and even antibodies against these proteins. Generally the academic scientific research community sharply criticised these patent activities and particularly the Human Genome Organisation expressed concern that the approval of such patents may harm the exchange of scientific data and may impede the transfer of important technology of benefit to human health.

As the NIH application is pending at the European Patent Office it is impossible for this contribution to indicate how it will be examined. It can be said now that in addition to the usual requirements for patentability, such as novelty and inventive step, and in addition to the requirement of unity of invention a key question for assessing patentability will be whether the ESTs solve a technical problem. If the technical problem underlying the alleged invention is to be considered as the provision of human genes probed by the ESTs but these genes have not yet been provided and have unknown functions, it could be argued that the invention as disclosed does not solve this problem.

If the expectations of the academic scientific community and particularly the Human Genome Organisation do not become true and NIH and the Medical Research Center in England choose to pursue their cases further, in the near future it will be necessary for the European Patent Office to consider whether or not this type of patent applications will be acceptable under the stipulations of the EPC.

Discussion: Session 3

Vogel: I thank Dr. Sahai for reading Prof. Swaminathan's manuscript. In my opinion, it is an excellent contribution because he is looking at the problem from a systems approach. He not only takes into account one line of thinking, he considers the entire system including also the social and the population aspect and the aspect of international cooperation. And I must really say, it is very unfortunate that Prof. Swaminathan was not able to be here because we would have liked to discuss all these things with him. We had a very interesting block of contributions this afternoon and we can now start discussing them.

Garthoff: I would like to comment on some of the aspects that Dr. Kräußlich mentioned in the beginning. I think there was one issue which should be a little bit elaborated upon. There is one specific aspect in transgenic animals that we should approach somewhat further. That is the issue of having transgenic animals for production of pharmaceutics. There is, for example, a sheep that is producing one kind of a substance which, up to now, is extracted from human plasma and that is used for a population – subpopulation – of about 100,000 people world-wide. If it were not for the aspect of having a chance to patent that kind of a transgenic animal, I'm quite sure we wouldn't have a chance in the future to supply all these 100,000 people world-wide with that one product they need that is α_1Pi or α_1-antitrypsin. So that Scottish company coming up with a transgenic sheep that produces this compound in milk and in collaboration with our company distributes it world-wide, is apparently only doing this because of the possibility to have a patent for it. So, a patent on a transgenic animal helps to produce a compound that none of the companies would have produced otherwise. I think that's an essential issue which enforces the aspect that was mentioned this morning by Prof. Straus: Patents enforce human health.

Vogel: I shall ask you a very stupid question – it's very stupid because I should know it from my background. Of course, I do know α_1-antitrypsin deficiency; it is a genetic defect leading to abnormalities of the lung and the liver. I have not heard, so far, how α_1-antitrypsin could be used to therapy. I do know that somatic gene therapy is being prepared by some people, but is it actually being used for therapy? Or is it used for completely different purposes?

Garthoff: No. Those people at the moment being afflicted with that kind of deficiency may get it from our company which is extracting it from plasma. The only problem is, it's such a tiny amount in plasma – even world-wide plasma substitutes wouldn't help us in producing enough for those people; so, in the end, it is almost a lethal disease, as you know. Those people are heavily dependent upon that one tiny amount in plasma of other people.

Kräußlich: We have two problems. One problem is expression, a super expression, as has been accomplished in plants, it is detrimental in animals, low expression, it is not economic, only normal expression is desirable. The second problem is purification of the product. In Germany products of transgenic animals are not considered as normal animal products. The costs for the industry are increasing with increasing demands.

Sahai: I have a question or I actually seek clarification from both Mr. Lange and Mr. Kräußlich. This whole question of breeders's privilege, research exemption, I'm not terribly clear about that and I would like to be because in my mind, I think if you have just a research exemption that allows you to fool around with the material in the laboratory, but you cannot commercially use it, then it amounts to a hill of beans. It doesn't really mean anything at all. So, I would just like to clarify – is what is permitted, for example in the revised UPOV – is that only a research exemption or does that extend to commercial application, as well?

Lange: I spoke about the breeders' privilege and I said before that we would like to differentiate between research exemption in patent law and breeders' privilege and the UPOV system. And even under the revised text of UPOV there still exists this breeders' privilege which means that you can use protected material of a plant breeders' variety and use it, for instance, in a cross and develop a new variety by using this other protected variety. That is free also if you market the end product, if you don't fall under the new concept of essentially-derived variety.

Folman (Israel): I would like to ask Prof. Kräußlich a question. I think in transgenic animals we have a special problem, for instance, with microinjection. It depends very much on the specific success of the process, what sort of animal you are getting. You may get, with a certain gene, absolutely different types of animals or different clones. And if you give a certain company a patent for a transgenic animal for a certain gene, to produce a certain compound; and next year or next month there may be another company which will have a much better animal which produces much more of this compound and it will do it in a different way which is much more efficient and productive. What are you going to do with your patent then? Will you give another patent, will you say that the previous patent is not good enough? What's going to happen really from the legal point of view?

Kräußlich: I can't answer this question from the legal point of view. Different results, in primary transgenics, selection between transgenic animals and propagation of desired transgenics is the breading procedure for transgenic lines. That's in principle the same procedure as in normal animal breeding. The economic exploitation of patented genes in transgenic animals requires two steps, first establishing primary transgenics, second selection, evaluation, production.

Vogel: I want to follow up this question a little bit more because this is what bothers a biologist most in the entire story. Of course, I have learned from Dr. Straus, and especially from Dr. Beier that repeatability is a very essential part of patenting, which means that a procedure or an invention, in order to be patented, has to be repeated. And this is not the case with transgenic animals at all because insertion of the genetic material, insertion of the gene, may occur or may not occur and may occur in quite different parts of the genome. Depending on the promoter, for example, where it happens to be inserted, this may lead to quite different varieties. And my question is now to Dr. Beier, is it enough in this situation when the animal or whatever is deposited somewhere so that everybody can control it?

Kräußlich: May I say something to the problem of repeatability. I think it is a thing on two stages. If you have primary transgenic animals with the same gene construct, you may have different results. But if you have selected one improved transgenic animal, you have a complete repeatability. So, we have many transgenic lines in our laboratory and we have quite a normal Mendelian segregation and if you have homozygous animals, it's completely repeatable. I think we have to differentiate between these two stages of making transgenic animals.

Vogel: This means, you first make the mice or the other animals and then you select the best copies or the best specimens – the best individuals – and now you start breeding with them.

Beier: I can provide the answer. You know that in German case law the "red dove" decision has accepted in principle the patentability of biological inventions, but has denied in that very case the patentability of the red dove because the court says the breeding process could not be repeated with the same results and will in practice not be repeated, so one of the basic conditions of patentability is not fulfilled. This was then smoothed down because now the general opinion is that if, for instance, the breeding procedure or another production procedure could not be repeated or if the production of the biological product could not be described in the patent specifications in so much detail that in every case the same result occurs, then this insufficient description could be completed by deposition of the plant or other biological material so that third parties have the possibility in reading the specification and having access to the deposited material can get access to the information which is indeed a prerequisite of each patent protection. But I would also add some explanation as to the difference between research exemption and what Dr. Lange rightly doesn't call research exemption. I would rather call it a free ride for plant breeders to use the material which other breeders have produced and I think it can only be explained in the context of the former breeding methods. These methods, I think, needed rather free access to other plant material, even if it would be protected by breeders' rights since the qualities of the product which is used for selection and cross-breeding was not so important in relation to the breeding process. But now, if you use the modern techniques in the plant breeding process to come to a new variety of plants and you say, it does not cost so much, then a research exemption, as we have in the patent law, is the better solution. And the difference is, indeed, you can do everything for research purposes. At my institute a dissertation was produced dealing with this topic. It will be published this year and also takes into account the American and English case law in that respect. It is a difficult delineation. Everything is free which is concentrated on the object of patented inventions. A patented invention can be used by everybody, not only by universities and non-profit research organisations, but also by industry if the patented invention of a third party is used in order to implement the invention, to further develop, to circumvent even because the circumvention of an invention is absolutely in the sense of the patent system. This is part of the technological progress. But if then the result is come out from these research activities and one firm or the university will start to commercialize the products which are a result of that research, then the patent system again comes into function and then the owner of the ameliorated invention must deal with the original patent owner. If he has found a better solution or a more economical production method, I am sure that the original patent owner will give the license, and in general will get a cross-license, because he cannot market his old invention when he produces it three or four times as expensive as the other. So, I think this is the mechanism, but I agree that the research exemption we have in our European patent law which is restricted to experiments – experimental use – should be enlarged, to allow every serious academic or industrial research which is concentrated on the object of the patented invention. I think, maybe the courts can do something, but here the legislator is called upon to clarify the scope of that exemption.

Vogel: I have another question for Dr. Beier immediately connected with this which has been really bothering me. Go back to the example you have given on α_1-antitrypsin. This, of course, is being produced by some farming, as you say, and you treat certain patients. In order to really find out whether it helps or to which degree it helps or when it helps, you have to do research on those patients. On the other hand, those patients, of course, are coming to the hospital, the expenses for their treatment are being met by, let's say, health insurance systems, or they have to pay for themselves. But you do research on them because you have to find out whether your product actually does what you expect from it. Does this go under "research exemptions" or doesn't it? Is it commercial use?

Beier: I think I must pass the answer to Mr. Straus and Mr. Moufang because I think they have been involved in cases like that. It is a very critical field. Clinical experiments either fall under the research exception or not. Our supreme court has said, if the clinical experiments are deemed to get the approval by the health administration then it is in the commercial sphere and it could not be done without the consent of the patent owner. On the other hand, one could come to the opinion that if these clinical experiments are in the state of research, then they may fall under this exemption.

Sahai: I want to go back to the clause of licensing that you mentioned when you were trying to explain this. And you said that in the case of somebody else producing it more economically, you would imagine that the original patent holder would almost certainly grant the license. Why would he do that? If he has the monopoly he doesn't have to give the monopoly away, he can produce at four times the cost. If this was the only thing on the market, why would he want go give away that advantage?

Beier: I can answer that very quickly. He can produce, for instance, the product according to a non-patented procedure. But if the third party has invented a better procedure for producing or he has found out a new use, a second medical use which is much more interesting, for instance, take aspirin. Aspirin we have taken for years and years only for headaches and then somebody found out that it could be used in a very, very little quantity to prevent coronary heart disease. This would be an invention if the new use or the new procedure is new and non-obvious and then the old patent owner has the monopoly on this product. But he cannot use the production of aspirin for this new use and the owner of the new invention could not produce it without the consent of the original patent owner. This is the basis for cross-licensing and if it does not function, then one could imagine something which is already introduced in the legislation of some countries, that is a special dependency license.

Straus: Perhaps I should follow the order of questions and remarks. The first was the question concerning animals. What to do when you have a transgenic animal and you have to make some additional selecting steps? This is a very good example to show the possible sharing of the work between the patent system and let's say, a special system for protecting new varieties, even of animals. The patent system really protects the solution, the technical instruction how to make that transgenic animal, how to construct the new vector, how to bring it into the fertilized egg. Then you have *the* product of that invention, the animal. Now if you need additional steps and you have a patent on that animal according, at least, to the German situation, there is no requirement to reproduce the animal or any other biological entity in genetically identical form because until now we really don't know where the transferred gene is located. It's a phenotypical identity which is required and sufficient. With the invention of conservation methods there are, not only the frozen embryos, but also other ways, e.g. the deposit of the gene constructs to get the repeatibility of that invention into place. The second step of this selection could be viewed either as a new invention possibly because it might be necessary to involve an inventive step to find out which individual is that who will really inherit that trait. Although it is possibile in principle to get a new patent, in the normal case, as it is now in plant breeding, these selection methods are something which is under the auspecies of the plant variety rights' protection. Here the protection is linked only to the specific variety derived from that individual. This is one explanation. As to the second explanation regarding the clinical trials: I've written and published quite an extensive paper on that and I may just mention that the opinion expressed in that paper was apparently successful. I had to deal with the issue of whether experiments or clinical trials involving subject matter protected with a dependent patent can be performed. This is a very complicated issue, but can be relatively simply explained and was the basis also for a few remarks I made this morning. There is one patent on a substance and this substance has a less favorable biological activity because the constructs, the vectors and the host organisms are different. But since the substance, the DNA sequence as such is protected, everybody who uses this sequence is now infringing the patent. But since he's using different vectors, different 5' coding regions in that construct, his product apparently is a better one and this is something which is useful for patients. How can the inventor find out whether his or her product is really better or not? He has to perform that clinical trial. My contribution was published in a Festschrift for a retiring presiding judge of our Supreme Court. He wrote a nice letter – stating inter alia that there were reasons to rethink the whole problem. Maybe we will get a new case law – I don't know – but I hope so. Also the chemical industry is split on that very issue. But they are once on the one side and the next time maybe on the other.

Herbeck (München): I have a question for the plant breeder, and for a patent specialist like Prof. Beier. Modern plant patents cover usually, if you read them, plants, parts of plants and seeds. Now, look at, for example, cotton, which is closest to the market in the States, cotton which is resistant to a certain pest. Does the patent cover the seeds if they are sold? And then, in addition, the sale of the cotton which is sold to the consumer or is it just one thing that is covered by the patent?

Beier: Well, I'm not a patent lawyer and, therefore, I would like to give the microphone to Mr. Straus. He can better explain the situation – if you referred to the situation in the United States and the Plant Patent Act, for instance, or which patent do you mean?

Herbeck: Well, I don't know if there's so much difference between the United States' patent system and the European patent system, in this regard.

Lange: Yes, I think so. As I have been informed, in the United States you have three possibilities to protect such inventions. You have the Plant Patent Act of 1930 which covers all asexually reproducible plant material and then you have the Plant Variety Protection Act under the UPOV system and then you have, of course, also the Utility Patent. The three systems are quite different.

Rowland: I am a patent attorney. In talking about a patent, you have to see what is claimed and so we are allowed to claim in utility patents every aspect that you mentioned including the cotton to the extent that the gene that you've introduced is also present in the cotton. So, the real question will be, and I will try to discuss it tomorrow, what the patent office is willing to grant you. And taking the example of the animal, they could say that you have to go to the F_2 species and have to prove where that gene is as far as the chromosome is concerned and they'll give you a patent for that animal with that integration because it works and you haven't shown that this gene in a different chromosome would also work. We are now moving toward very important questions of scope, that whether we will be given something quite broadly, as you described it, any plant with that gene anywhere and any aspect of that plant or will we just be given the plant you have shown in an F_2 variety and we will have to show where that gene is in the chromosome of that plant. And then the question is, will I invest money to make that one species for my own use and I control only my own product, but my competitors can make other plants which will be as good, better or poorer, but they'll be able to use what I have taught and led. And the answer to that question is now being developed at least in the United States and I would say that our office is getting tighter and tighter about what they're willing to give and how much you have to prove.

Sahai: I think the cotton that he's talking about is the bollworm resistance in cotton for which Monsanto has a utility patent. So, that extends to the cotton variety and its seeds.

Sela: We do have a utility patent on a major trait which a floral mechanism in strawberries. My question is whether we can protect it just by using the utility patent or do we need to register and protect each one of the ten or so varieties which share or include this major trait?

Straus: I think the best way to protect traits is by protecting them by utility patents because you have a very broad protection in all material which is produced by this procedure and I think, in this case, it is not necessary to protect also all the different plant varieties. In order to get a plant variety protection for strawberries you have to first look whether they are in the list of protected taxa of the country where you would like to sell that variety and in many, many instances this is a fairly tricky issue. Some countries with good breeding activities in one area offer protection for their varieties but not vor varieties primarily bred in foreign countries. And since the UPOV type of protection does not extend to the final product and its marketing, plant varieties not included in the list of protected taxa in country A (e.g. Holland) can not only be freely propagated in that country, but the end products of such propagation, e.g. ornamentals or fruit trees, can also be exported into country B (e.g. U.K.) to be planted in the gardens and orchards in that country, although these varieties are protected there. This was completely legal and I mean, one should really realize and understand what the UPOV type of protection does and is. It will be, of course, improved after the entering of the force of the next act. Then also the importation will be prohibited, but until now, this is a kind of a patchwork. And since a lot has been said about the farmers' privilege and about the breeders' exemption, I must add that there is also a very large breeding activity in the field of hybrids. And there is no breeders' exemption under the UPOV convention which would apply, because for the production of hybrids repeated use of parental lines, not covered by that exemption, is necessary. In addition to that, there is no farmers' privilege because usually the hybrids are infertile. So, I mean, we have a large field of agricultural activities living without both of those exemptions and, as far as I understand, this is not the worst case of breeding activities. This is just to make a few clarifications.

Folman: Just one word – these hybrids are not sterile, but they will not reproduce themselves. They are not sterile necessarily, but they will not reproduce the same variety.

Beier: Before the war between the breeders and the patent side there comes again into hearing, I would like to say that we had three weeks ago the annual meeting of our association and there was a workshop where the two speakers, Prof. Straus and Dr. Lange have spoken about the same subjects and I gave that workshop the title: "Patent and Plant Breeders' Protection. Peaceful Co-existence?" And then Dr. Lange spoke and he started saying that also the traditional breeders use very much the new techniques and they are also interested in patent protection. And, as I heard that, then I said: "If I had known that before, I would have crossed out the question mark." Indeed, I think there is a growing tendency for a peaceful co-existence between the two systems especially since the traditional breeders are using also the new techniques and are becoming aware of the advantages of patent protection. I have always been of the opinion that there is room for both systems. The patent system should protect new findings in biotechnology, new biological procedures if they are repeatable and controllable a.s.o. and they should also be open to protect plants with new properties like antiherbicide a.s.o. But with that invention, there is in general not a marketable product; producing the marketable product is the function of the traditional breeders. I think the division of labor between the two is good for society. With the growing introduction of these new biological techniques, this war between the patent defendents and plant breeders' rights defendents may indeed come to a peaceful co-existence.

Lange: I'll open the battle! I agree with you that the plant breeders need, also, a well-functioning protection system for the "Vorprodukt", I would say in German, for the products which have to be developed afterwards by classical plant breeding work to make out of this new plant varieties. I have said in my lecture that biotech inventions doesn't lead to new plant varieties, itself, that's not the normal case. You need further breeding work. And, therefore, a good co-existence between both systems is necessary. But, I must say, there is a real difference between breeders' exemption and the research exemption under patent law and we believe that we need to have the breeders' exemption in the form as it has been formulated for further breeding work. It is a big difference to the research exemption under patent law and even the case which Prof. Straus mentioned is not a real exemption from the breeders' privilege because it only deals with production of a hybrid and not breeding a hybrid. That's a difference and if you take hybrid for further breeding work, which is also possible today, not using the parental lines, but using only the hybrid itself and crossing with others, you are under the breeders' exemption. You don't need to produce repeatedly by using the hybrid.

Sahai: The distinction that you make between animal and animal varieties and plant and plant varieties I think has left quite a few scientists mystified and I seek clarification there.

Vogel: Of course, I must also confess that I really didn't know what an animal variety and what an animal species or what a group of animals a.s.o. is. I think there we need some explications.

Gugerell: I would not like to come with a definition of animal varieties. I could quote the definition of a plant variety from UPOV. Practice is that – this is also based on decisions from our boards – that if the invention is applicable to groups of plants or animals which are higher in the taxonomic ranking than animal or plant varieties, for example, families or something like that, then per se protection on plants and animals is allowed. If the invention and, of course, all these recombinant DNA inventions are applicable to larger groups, – however, if an invention were made just relating to one plant or animal variety, we would interpret such a claim as referring to that variety.

Sahai: I'm afraid it's still not entirely clear. I mean, did I understand it is an individual plant as against a family of plants that constitutes a variety?

Gugerell: No, a variety is something below a species. It's questionable whether it's still a taxonomic level but it's definitely below the taxonomic level of the species.

Sahai: So, a bunch of plants is a plant variety and *a* plant is a plant, is that it? What is a plant per se as opposed to a plant variety?

Gugerell: A plant variety definitely is a group of plants which fulfil certain requirements and this is defined in the definition of UPOV – stability etc. A plant per se, having the gene incorporated in its genome would be interpreted depending on how the invention is diclosed in the patent description. And if it's clear from the description that the invention is applicable to not only one variety but to a larger group of plants, then we would grant a patent and grant claims to plants per se.

Straus: I would just like to assist a little bit because we have this problem all the time under the higher taxonomic units. If you try to say – the whole family is, of course, a higher taxonomic unit – the impression might occur that you are protecting the whole taxonomic unit, but what is really protected are all plants carrying the new traits belonging to different species of the same family. This is possibly something which could clarify the problem. Prof. Schell from our institute in Cologne applied for such a patent. These are transgenic plants produced by using his Ti-plasmid method and being transformed in this way.

Vogel: Well now, Dr. Straus, I don't understand anything anymore. First we have said that a variety is a group of plants below the level of the species and now you say that a variety may even be a group of plants above the level of a species.

Straus: I was talking about the family and I am be free enough to claim at least for me that for a lawyer, but possibly also for some other professions, it is very difficult to understand all those specialties and I would be always very hesitant to instruct a surgeon as to how to make a special move in heart surgery. Maybe this is a part of a problem we do all have when we are discussing these topics.

Sela: I would like to make a comment which has nothing to do with the situation as it is, but again from the scientific point of view as a scientist. I think it's very dangerous to make definitions and forget definitions that have been made by people, by men, and can be changed by men. Now, once you make a definition, you are bound by it and you forget that you can change it and then the definition of what is a taxonomic level is going to change very quickly these days. I just was arguing with an entomologist a few months ago – and he defines the species of an insect under the microscope and I told him, look, in a few years you will define it by its DNA analysis and you'll find out that whatever you have defined as a species is, as a matter of fact 10,000 different species. And so I think all these definitions – a variety, a species, a family, as it comes to law and patents – should be forgotten and every case should be defined by whatever you are patenting or whatever you are protecting and you have to be able to define it in whatever agreeable way is probable by sequencing or other genetic make-up that you can provide and forget about terminology.

Sahai: Without going into the complexity of who should instruct a heart surgeon and whether a taxonomic definition is going to change, I would like to reiterate that I still remain perplexed as to what the distinction is between "a plant" and "a plant variety". If I understood the discussions of the proceeding sessions, some amount of this discussion has to do with the circumvention of Article 53b of the European Patent Convention. This is allowed but that is not allowed. I'm trying to understand how you bypass this rule as to what is allowed and what is not allowed. My understanding is – and correct me if I'm wrong – that some amount of that circumvention hangs on this particular distinction, and I think it's important to understand the way this distinction has been made. So, I'd appreciate light, so to speak, being shed on the subject.

Grunwald: I think I could try to answer it. It's under this regime argued that patent protection could cover in the field of botany anything not defined as plant variety in the sense of the UPOV. So, you have a clear cut distinction. The distinction in the EPC was tailored after the UPOV convention. Parts of plants are patentable, tissue cultures and plant cell lines – those are the matters.

Waugaman (USA): I am speaking as neither a scientist nor an attorney but as one who has, like Reinhard Grunwald, advised scientists for several years on the issues of patenting research results. As I was telling my colleague, Dr. Rowland, if you don't know the difference between a plant or a plant variety, or between an animal or an animal variety, call your lawyer. The issue becomes one of whether or not the discovery is important enough to seek a patent, and if it is, what kind of protection is going to give you and your commercial partner the greatest advantage? And you try to argue the best case with the best scientific information that the inventor can provide to secure the best patent protection.

Erselius (Munich): I don't think that the definition actually is the problem. It's not so important to know whether it's a variety or whether it's an animal or a plant per se. I think that the people just can't understand why a variety should not be protected, but the plant per se should be protected. It's a completely arbitrary definition.

Gugerell: The reason why plant varieties are excluded from patent protection is quite clear. The old UPOV convention contains the provision that protection under both systems is not possible and therefore, plant varieties have to be excluded from patent protection. Prof. Straus has already mentioned today, why animal varieties were also included is not so clear, but they are also there.

Lange: Well, I must say that I cannot share your opinion of the meaning of Article 2 of the old UPOV convention. This Article 2 forbids double protection, but I think what is meant there is only the relationship between a plant breeder's right and, I would say, a plant patent – not the relationship between plant breeders' rights and utility patents. In the score of UPOV, it was not possible to regulate, the relationship between patents and plant breeders' rights and therefore, now the double protection is given up under the revised text and that means, in my opinion, nothing.

Gugerell: It means nothing because we still have the exclusion in our convention. It only means that the patent side will be now free to remove that if our administrative council would agree to that. That's another question, but a least the UPOV people could not hinder us from removing it.

Vogel: I would like to ask a different question on a slightly different topic. And this is, of course, what concerns me most is the patenting of human genes. And you argued – let me repeat your argument, more or less – you argued that in Germany, genes may be patented despite the fact that it is of course not in itself an invention but a discovery. But it is only a part of an invention to produce a certain product. Well, so far so good. However, now it should be used for quite a different purpose. I took the example first, you use it for making a product, later on you use it, let's say, for somatic gene therapy. Then you argue because it is part of a complex that has been patented, it now, despite the fact that it is, of course, a discovery, acquires, let's say, the quality of an invention because it is part of an inventive activity and therefore, it is also protected for quite different purposes. And this argument, I must say, is too involved for my understanding. But could you please comment on this.

Gugerell: I would say that this gene, at least in our view, is not a discovery. That's how we see the problem – it is not a discovery because a new use which was unrecognized before has been attributed to that gene. After the invention has been made you know that the gene is useful for doing something. This was not known before. That makes an invention out of a discovery and therefore, it's not a discovery, it's an invention and then I have argued, if you get absolute product protection, any use is covered by it.

Vogel: I understand, of course, you repeated more or less what I had said before. First it is a part of an invention despite the fact that in itself it had to be discovered, but there was no use for it and therefore, the entire thing is an invention and therefore, the part of the invention which is certainly a discovery, in itself, will now be protected for quite a different invention and this is a logic, of course, which I cannot follow. The entire thing has a practical consequence and maybe Dr. Straus or Dr. Beier may say something on it. I would find things much more agreeable and I would be much more satisfied with everything if it would be possible to patent, – let's stay with this human gene because that's what interests me most – the human gene can be patented only for a special use, for example, the gene together with another procedure to produce interferon or something like this. But the gene as such remains free and the gene sequence, when somebody uses it for some quite different purpose, then it would be a new step and for this new step it is not protected by the original patent. I think this solution which would – well now I am arguing opportunistically – also be acceptable to the more benevolent part of the general public and would remove much of the inacceptable aspects from the entire complex. And this, of course, could be derived by simple logic from the general principles of patenting, namely that discoveries are not patentable and inventions are patentable. Do you see, from your legal point of view any possibility to argue more or less on these lines?

Gugerell: Well, I have problems with that argument primarily because we have absolute product protection and that's our system. The system, of course, could be changed, but at the moment we have it and again I would say you always come back and say that it's a discovery. As soon as the use has been found for it, it becomes an invention in our view.

Vogel: That's absolutely not true from the logical point of view of a simple medical doctor.

Beyreuther: I think our problem is that the border between basic and applied science has disappeared on the gene level. We all know that the genes hold the key to the product – to the building plan. A receptor can be used to isolate the drug and therefore, it really doesn't matter where the gene comes from. If you use a yeast gene and make it look identical to the human gene why not patent the human gene? And the other thing that we have to keep in mind is that in the past if you worked on the bat, you may have used a very exotic animal. Now if you isolate a bat gene that has its homolog in the human genome, you do something very important and there is no exotic genome science any more. That's the reason why I think we really have to define that there is no difference between a yeast gene and a human gene if you observe it in function, if you introduce mutations and make the yeast gene a human gene, there's no difference. There's no ethical issue on the human gene that is different from a yeast gene. And another point to Dr. Grunwald, I think that the destruction of academic tradition only hurts mediocre scientists, it has no effect on the leaders in their field. I think that's a very important issue and there's only the danger that the followers may be hurt. If you look back into physics or chemistry, I don't remember any example of patenting having hurt a scientist, they just looked at a different system or did something that was an invention.

Grunwald: Let me just answer the second part of your statement. I wouldn't dispute that advances in science have their price. But I would put a question mark behind the statement that we destroy only dead wood or that we destroy only academic procedures. To repeat: To do this kind of research or science you have to work like industry. You have to put things under lock, you have to talk to your people, instruct them every morning what they can and what they can't do, and do so with everybody who goes into the laboratory. And I'm not talking abstracts. I could tell you things that would be quite a problem because we got substances under such a research clause and somehow they were found somewhere else. I could give you many examples concerning the misuse of substances. This kind of orientation changes laboratory organisation and you, scientists in the life sciences, are proud of your decentralized organization. Under this regime, you will not have a decentralized organization. You'll have a central organisation, you'll have lawyers telling you what to do and not to do. And they'll tell you, "You can't publish this," although you know your runner up is 14 days behind you. He'll say, "I need three weeks to look into the patent issue; do you want to have a patent or do you want to publish?" So, I think we can't make it that easy to say it's only the destruction of old academic tradition. It's something else. It changes the whole science community, the interaction within the science community. It changes science towards a result orientation which, in my opinion, from an administrative point of view, could to some degree be highly welcome. But from the point of view of science there should be many question marks.

Just one remark to the problem whether the substance occurs in nature or not. I think we have to say that the theory of absolute product protection was not undisputed in Germany before it was enacted. We had the discussion whether the first use was to be protected and we have some discussion especially in the field of medical use. But the main difference is, the product we are about to protect here, is not a chemical product found by industry, but by evolution and you can't circumvent it for some time to come. So what you have is not a "normal" chemical component which you can find through other means, too. Naturally, and that was the message I tried to give, the changing of the substance, its manipulation is patentable: The process is patentable, but be very careful with patenting the product itself. It's another thing whether you go through yeast – it's a means of production. That's a man-made thing. If you can transport the message to say, well I need the yeast as a factory to produce a component which is exactly like the human chemical, I think you will convince people much more easily than by saying, well, this is part of the human genome. But we are about to discuss it and I think tomorrow will be a good opportunity to do so.

Hess: There are many, many, questions which occured to me when I listened to this morning's and this afternoon's discussions and there's one little point I wanted to mention. If you have a gene product which has two different properties, one in metabolism and one in structure in a completely different area, what are you going to do? You get a patent for one product property or for another product property? I don't want to give the detailed case. Some of you might know what I'm referring to. The other point of view I wanted to bring up has a little bit to do with philosophy, what Dr. Beyreuther just said in terms of, let's say LDH from yeast or LDH from humans, what is the difference? Why do you identify a gene which expresses in the human liver with the label "human". The human genome has a label which I think is appropriate because if man is defined in a certain way, at least with respect to its makroscopic structure, by the total structure of a genome. But we are talking here and in all the other cases about little bits of DNA and if the source comes from plants or does come from a bacteria or from humans, then we give it an etiquette which is not appropriate. I agree with Dr. Beyreuther on this point of view. We should be cautious in applying certain identifications with respect to the source from which that DNA or RNA comes from. I think when we talk in terms of biology, then, of course, it's only the genome which defines a certain species. Perhaps the difference between the yeast cell and the mouse is only a tiny part of the structure, perhaps 10,000 nucleotides in sequence. To go a little bit further, I remember somebody once saying in one of those meetings I attended, what are we going to do if the difference between man and mice is only a tiny stretch on the total genome? It might be not so pleasant for our personal feelings. I only wanted to point to the problem that we should be more cautious to use the etiquette "human" with respect to small molecules. Nobody certainly ever cared about whether nicotinamide comes from humans or from plants or whatever. The source is not involved in the whole problem.

Grunwald: Talking about human origin of genomic material, I think is a very important question and especially before tomorrow: You'll have to convince laymen, you don't have to convince me. I'm quasi an interpreter between you and lay people and parts of the public. And I can only tell you that the notion that *humans can't be patented* is a definite notion in the public mind, and let me add, in my mind, too. You can't patent humans. So, you have to transport the differentiation. For you as experts, it's clear and I know, too, that the genome of the chimpanzee is well over 90 % the same as ours. But people on the street who have to vote don't know that. And this is a bitter truth. As I said before, you experts went out on a limb and now everybody sees where you already are and they are afraid. The onco-mouse decision and the reaction to it showed that. Please don't lie into your own pockets. See the problem. It's not only cutting off old academic tails. It's much more. It has something to do with how man sees himself. Whether he sees himself as the sum of many chemical components or whether he sees himself in a different way. I can tell you we had this discussion in our center and these were exactly the notions and the fears which came up and you don't overcome those fears by ever louder and ever more clearly saying, this is nonsense.

Vogel: I think many of the objections could be removed if we would go away from this entire object protection. If you could actually go into protecting certain DNA sequences only for certain purposes. And I think if this could actually be done as a kind of rule, it would not be important whether you take the same thing from the yeast or from maybe a chimpanzee or from human beings to make the same thing; then it is only a matter of practicability – where it is cheapest and where it can be gotten most easily. It would not in principle be different from, let's say, insulin therapy, whether you take insulin from the pig or from human beings, a.s.o. One may be a little bit better than the other, but the problem involved is actually the protection of the entire gene and we can see in the future that not only a few genes will be protected, but all genes or many of the genes will be protected by patenting law and then the entire genome research in humans is hampered by these complicated systems of patent protection mainly because most of the research is carried out on patients. This means we pay for it or the health insurance pays. I am very glad to hear that Dr. Straus has clarified this point partially for clinical research. However, it really remains a problem. I'm arguing not so much from a philosophical point of view. If you would tell me that I am not very much different from a chimpanzee, then I would, of course, agree with you as a biologist. But still, if you could eliminate this point, I think that the acceptance in the general public would be much better. What you said and also what Dr. Beyreuther said and what all these patenting people say, disregards completely the reaction of the general public on what we are doing. And you may look at it from any angle you want to, we are not regarded as slightly more developed chimpanzees, we are regarded as human beings and this is very much justified. Let's take the genes of Eshericia coli – I'm just going down the taxonomic ladder – we have many genes, look at the galactose metabolism, which are identical in E. coli and in human beings, more or less. And it requires a degree of rationality to accept this, which you cannot expect. As a scientist I can be so abstract sometimes that I can accept this, but not as a human being.

Hess: I don't want to be misunderstood. I tried to focus on one essential point and this point was, I am hesitant to designate a certain gene to the source to such an extent as it is usually done because of the argument Konrad Beyreuther also brought about: I do not see, if I take a certain gene for a certain enzyme or protein from humans, what the human property in this gene is. After all, the same gene with a slightly different sequence perhaps might come up in many other species and we know this, of course, from energy metabolism where the unique identity of many essential properties are obvious. Now, the other point of view is the term "human." I would rather restrict it to the totality of the genome and I would even go one step further: One genome can only be described in terms of the individual of that genome which it is carrying and not for mankind. I wanted to bring to our attention that we have to think in much more anthropological terms when we define these types of structures which are implied in this immense complexity that we now call "genome." I would distinguish quite clearly between the properties of a gene, on the one hand, and the genome, on the other, and that might be much more helpful in the future, also with respect to the difficult problem of species and subspecies a.s.o.

Garthoff: I just would like to come back to the same issue. One of the issues I think we're mixing up here is, on the one hand, the economic aspects of patents and the other aspect of ethics with regard to the human genome and one of the issues that I think is very helpful is differenciating the human genome, on the one hand, which is the complexity of human beings, and the one gene which is in patent terms just a product. Let's put it down to terms. This is a product and a product can be patented.

Vogel: It's not a product, it is a discovery because it is there, it is not produced.

Garthoff: I'm sorry to say it's similar to any other molecule. It's several parts of a molecule, but it's simply a molecule, it's like Aspirin®. We had this example this morning. So, if you come in with an Aspirin®, you have to use it. Well, you started out with a headache and later on you came in with another indication which also is helpful and patentable. It's the same thing with, for example, let's say we have a gene sequence for factor VIII. I'm going to produce a protein, factor VIII, that is going to help the hemophilic population. The next step is to use that same issue for gene therapy and it's exactly what, for example, our company is doing. So, the next step will be gene therapy. It's still a product from a patent issue. As soon as you come in and say, but when I have all those patented gene sequences, that's a whole genome and then it's man and then you're right in the ethical discussion – that's the issue – but for the gene sequence, by itself, it's not a difference, there is no difference between dog and man with regard to that.

Hess: Let us assume you discover a protein, and later on you go to the DNA sequence. The first time one publishes, I would think, this is a discovery, but what happens later on, 10 years later or two years later and whether it's used for any purpose, that might be an invention. I would like to make this clear distinction. We are – Konrad Beyreuther might refer to it – in a queer situtíon, because in biology and certainly in molecular biology and biochemistry, we have not been used to the current situation that any newly discovered natural structure and function was immediately brought to the next patent office. We are not used to this situation and we somehow have to deal with it because, up to now still a large majority of my colleagues never think about it, they never do it, but there are certain groups in our community which do it for one reason or the other. There is a change in, let's say, general habits. What this is due to, Herr Grunwald, I don't know, but sometimes I have the feeling that this has something to do with the very fact that nowadays politicians are asking in many countries for immediate reward and they have not the patience and even the time to wait, let's say, the 20 years until this might be rewarding for the public.

Grunwald: Objection your honor. To blame the politicians is only part of the truth, and in this present discussion we should not look too much at other influences which have the same result. You're completely right – one effect blends into the other. It's very difficult to tell our American friends that our Minister of Research and Technology has been "promoted" to Minister of Public Transport and that now his budget is more than four times as large as it was before. So let's not go into this, let's stay within science. And the science aspect we discussed here was that the chemistry of life is a product. In my opinion, the problem would be much easier to tackle, you're completely right, if we look at it as a product. In German patent law, what is protected is technical teaching. So, where is the technical teaching? In the procedure, you have technical teaching, but what you find is nothing else but a discovery – and in 100 years we will be looked upon as the man with the flintstone knife cutting his steak from a mammoth: Had patent law been in existence already, you could have patented the knife, but not the steak itself – unless you changed it and grilled it, for example. It's the same thing with isolation gene particles.

Folman: I think actually what we have here, if we look from the historical perspective that different countries have patent laws for many decades and centuries and now they encounter a new situation where the human genome or part of it or DNA sequencing comes all of a sudden into commercial interest and you have to deal with it. The reason you deal with it is because there is a sudden commercial interest. Somebody is willing to pay a lot of money, somebody's investing and probably what happens is that you take a certain sequence which is a discovery, it has a certain use and then you say it's an invention and because for many other things like chemicals and mechanisation you had certain rules, you apply those rules to this sequence without having perhaps the ability to change it. What should have really happened is, the human genome and the DNA sequencing perhaps require an absolute new set of rules with regard to patenting. The old set of rules do not apply in many cases. So, for instance, should you be able to say that a certain sequence is giving us a certain product – a protein or insulin or whatever and you give the patent only for this and not for anything else, then you would be O.K. but then you are against the old patent laws. So, the legislature, the politicians, should be able to differenciate between the needs of biological patents like DNA sequencing and the previous needs of chemicals and other things. They have different requirements and you cannot very easily put those new things into the old laws. Probably we need new laws for those new biological inventions.

Vogel: You're absolutely right, I would say these problems can actually not be solved by the European Patent Office.

Gugerell: But if you say that even here in this room we don't have consent, how could you expect politicians to have it?

Straus: If I may come to the more trivial fundamentals of the patent law once again. I would first like to try to explain why something is an invention and not a discovery. You say there is no teaching when you have a product invention. And may I start with that. Every patent applicant has to describe the way how he did it and how this way, the teaching, can be repeated at will. In the case of DNA sequences applicants spend quite a lot of space on describing how they did it. And what one should bear in mind is that if the way of discovering or finding it, the teaching is already trivial, then no patents can be granted irrespective of whether this is a DNA sequence or whatever. If I may say just a sentence of criticism: Maybe the patent offices are too liberal in applying this specific requirement of inventiveness. This is difficult to estimate, but sometimes one could get this impression. This rule on the distinction between discovery and invention is a very old one. We have had these problems for 60 or 70 years and they have been solved in the way we are talking about here. What about Prof. Vogel's idea about the purpose-bound product protection? Unfortunately, our boss, Prof. Beier left. He was a pioneer fighting for this kind of product protection years ago. It turned out that he lost the battle, whether wrongly or rightly, I will not try to estimate. In any case the present situation is, as explained by Dr. Gugerell, but I would be cautions in saying that this causes a horrible problem. For the time being, this statement can really not be made. I was free enough to use the *cephalosporin* example this morning and one could add others. Other inventors were able to find other *cephalosporins* and the same occured in the case of cortisone and many,

many, others. We have patents in this field already for some time. The interferon application was submitted at the beginning of the 80s. So we don't have any serious problem until now and in about 10 or six or seven years the patent will expire. Before trying to change the law along these lines, one should really have a concrete case, not just an abstract one. I am talking about substances, not about microorganisms. There is no evident problem until now. If that would appear we have some other remedies first to be applied, maybe a compulsory license, although a part of the industry wouldn't be very satisfied with this kind of solution; but we have this remedy and this should also be applied before making a big plead to change the law. And if that wouldn't work our neighbors – I was hesitant to say that this morning – our neighboring countries Switzerland, Holland, the UK, Denmark, France – they have all dependency licences and everybody can deal with that. They have different requirements for the dependency licenses from more or less nothing in Holland and some very reasonable requirements, for instance, in Switzerland. So there would be some intermediary solutions before one should go and ask for new laws because to change the laws is much more difficult than to discover a new substance or even to discover a new DNA sequence coding for something specific. Maybe a last remark, with respect to what has been said by Prof. Hess. Of course everything is true. What bothers me is the point of attachments. If the DNA sequence of a human origin is a unique one, the unique one coding for a specific protein which is of essential importance for humans, then there might be – I wouldn't say in philosophical terms – but in terms of patent law with its blocking effects – a difference and then – like a governmental Swedish group which pleaded for it – a purpose-bound product protection could be discussed.

Sela: I just would like again to say one sentence as a kind of devil's advocate. All of us and especially the people of the legal profession should be ahead of events and not following the events and I am going to give an example which is very visible. One of these days we'll come up with a synthetic gene. Is it a product, is it a sequence, is it the sequence and the product together and if it does something, the products of the product and if you don't think of it now, we are in a mess.

Session 4

Patenting of Human Genes and Living Organisms: An Ethical Point of View*

W. Ch. Zimmerli

It is well known that questions concerning living organisms are of extraordinary interest to almost everybody. It therefore goes without saying that the issue of patenting living organisms unleashed (and is continuing to unleash) "a heated and somewhat confusing debate" (Moufang 1989) in the United States and in Europe during the late 1980s. This debate focused on legal and ethical questions, but, unlike other comparable cases, the legal and the ethical aspects of patenting living organisms and/or human genes cannot be completely separated from each other: the relevant pieces of legislation, e.g., the European Patent Convention (EPC), explicitly request ethical considerations regarding the morality of publishing or exploiting certain inventions. Article 53 EPC dealing with "exceptions to patentability" reads as follows:

European patents shall not be granted in respect of:

(a) inventions the publication or exploitation of which would be contrary to *'ordre public'* or morality, provided that the exploitation shall not be deemed to be so contrary merely because it is prohibited by law or regulation in some or all of the Contracting States;

(b) plant or animal varieties or essentially biological processes for the production of plants or animals; this provision does not apply to microbiological processes or the products thereof.

A substantial portion of the aforementioned ethical debate has, therefore, been directed towards this question of *'ordre public'* or the morality of publication or exploitation. Quite considerable portions of the objections raised against the patentability of life did not, however, dispute the morality and/or adequacy with regard to the *'ordre public'* of certain given cases, but of patenting life as such. In the following, I must thus first of all deal with this fundamentalist criticism (Sect. I). After that, I should like to take a look at the ethical discussion on molecular biology, gene technology, and genetic engineering methods in general (Sect. II), and then go on to outline a specific model of problem-oriented procedural ethics (Sect. III); by using this model, it will be possible to go some way towards solving the *'ordre public'* and morality questions with regard to patenting human genes and living organisms (Sect. IV).

---

* Text of the contribution partially received by the author for publication.

I.

The 'fundamentalist' or (as some – misleadingly – call it) 'metaphysical' (OTA 1990, p. 130) objections to patenting living organisms and human genes as a rule make use of some kind of sanctity of life argument, demanding a moratorium or something close to it. In particular it is argued that patenting of life violates the sacredness of God's own creation, thus transgressing the limits to hubris.

As a matter of fact, there may easily be something to these arguments, and I am not criticizing them – except for the fact that they are missing the point: they do not constitute objections to patenting, but rather to breeding, using genetic engineering methods, or creating transgenic plants and animals, etc. In other words, the mistake is that innovations in plant breeding and animal production are considered to be *inventions*, but this would, strictly speaking, not be an ethical argument. To put it in a nutshell, if one wants to raise objections, one has the choice of *either* objecting to new methods in plant breeding, animal production, and genetics, *or* of objecting to protecting individual (or collective) intellectual property. This, then, is the argument everything boils down to:

(1)(i) If something like legitimately acquired individual (or collective) property ("property *de iure*") exists at all, and
(ii) if the cases in question are cases of individual (or collective) intellectual property,
(iii) then they are morally legitimate and therefore to be protected.

We must, therefore, thoroughly examine conditions (i) and (ii) in order to be able to decide whether conclusion (iii) does actually follow.

With regard to condition (i), the history of the different attempts to justify individual (or collective) property ("property *de iure*") goes back at least as far as to Hugo Grotius and John Locke. Since they developed the canonic justification of property by labour in the seventeenth century, different versions of legitimizing individual (or collective) property rights have been put forward (cf. Carter 1989). The socialist criticism of the narrow conception of property *rights* has made it obvious that these are to be supplemented by the idea of property *duties*. In addition to the *social duties* already constitutionally guaranteed, one will have to expect *ecological* and *cultural* obligations to complete the justification of individual (or collective) property rights with corresponding property duties (Zimmerli 1992). This could again be one of the points where ethical questions of '*ordre public*' and morality could be raised with respect to the patentability of genetically altered living organisms and human genes. In principle we have to admit that nowadays individual (or collective) property has to be considered legitimate.

We can say with respect to condition (ii) that, as all relevant scholarly writings testify, the very idea of a patent as a temporary monopoly on the making, use or sale of products or procedures is closely connected to the idea of intellectual property. Although there is evidence of a few patents in the thirteenth century and although the Venetian Senate passed the first general patent law as early as

March 10, 1474, it was the concept of individual authorship that was needed to create the notion of a patent as an expression of an intangible intellectual property to be protected by law (Long 1991, pp. 875 ff.). It is not surprising, therefore, that this development is accompanied by highly controversial disputes concerning who first invented or discovered something (Merton 1957). With respect to the applicability of this concept of patent to living organisms and genes, there has been some disagreement on the question of inventiveness, and it has been argued that genes and/or living organisms cannot be strictly said to have been *invented* by somebody (cf. Koechlin 1993). At second glance, however, this objection of course turns out to be due to a lack of clarity with respect to the notions applied: Every scientific discovery, if made technologically applicable, becomes an invention. The cases in question, therefore, are inventions and consequentially examples of intellectual property of the kind to be protected by patent law.

For conclusion (iii) we can therefore state that, since they are patentable 'untangible' intellectual property, genetically altered living organisms and human genes are thus in principle morally legitimate. Q.e.d. It goes without saying that this is not a general ethical absolution for every single case of patent application concerning living organisms and human genes, but rather a counterargument against fundamentalist (or metaphysical) objections. As I have pointed out in my commentary regarding condition (i), the 'in principle' justification of individual and/or collective property nowadays is valid if and only if the property right is met by the respective social, ecological and cultural property duties.

With respect to the sanctity of life argument, however, we have to consider yet another aspect. Sometimes it is said that even if the patent that protects the intellectual property in the invention of technological procedures to genetically altered living organisms were morally sound, it would still hold that life is not subject to individual ownership. Again, the meaning of this objection must be thoroughly analyzed. The implication is either that the owner of a patent on procedures and/or products owns the products produced by these procedures, or that 'life' and 'living organisms' have the same meaning. Since neither of these implications is true, we have to scrutinize the following argument:

(2)(i) If a patent does not entitle its holder to private ownership of the products produced by the patented procedures but just to exclude others from its use unless they pay fees, and
(ii) if 'life' and 'living organisms' mean something different,
(iii) then to hold a patent to produce living organisms does not entitle its holder to privately own either the living organisms or life, but just to exclude others from using it unless they pay fees.

In this case the examination may be brief, as both conditions mentioned are self-evident.

For condition (i) we can say that as a monopoly by definition is the right to exclude others from certain transactions, a patent being a monopoly is so too.

For condition (ii), it is obvious that categories would be wrongly assigned (Ryle 1949, p. 20 f.) if one were to identify 'living organisms' with 'life'. What we call 'life' is the first-order quality which makes living organisms live; or in other words, whereas the specific term 'living organisms' may be defined extensionally as 'all beings that are organisms and alive', 'life' is the respective generic term which is to be defined intensionally only. This can be demonstrated by a simple empirical argument: due to metabolism, life can be preserved by sacrifying living organisms only; if an organism of higher order o_2 wants to 'stay alive', it is by necessity forced to eat and drink or to do something equivalent. This again is not possible except by profiting by the death of a least one lower-order organism o_1. Furthermore, it is not only possibly but – with one decisive exception (!) – it is not even sound either morally or legally to own living organisms, i.e., animals, plants, and microorganisms. The exceptions are, of course, human beings. As can be demonstrated with respect to the legitimation of individual and/or collective property by labour [see condition (1i) above], one of its crucial presuppositions is that the only legitimate owner of human beings can be God, who in his unlimited kindness presents man with the private property of his or her own body and the labour this body is capable of performing.

With regard to conclusion (iii) then, from this follows that – again 'in principle' – no violation of sanctity of life is connected wih granting a patent on living organisms and genes to an individual and/or collective person. Q.e.d.

II.

Having thus dealt with fundamentalist objections to patenting of human genes and living organisms to enter a rationale of the morality of the rights of patenting. Before we do that, however, it is necessary to scrutinize today's situation with respect to both the process of technologization and the changes that have taken place within ethics.

Our present time is characterized by an ongoing process of technologization which for its part is marked by the following conditions:

1. New insights can only be obtained using highly developed technical apparatus and aided by equally elaborate technical knowledge or skills.
2. The knowledge thus gained is not purely theoretical knowledge; rather often it is knowledge of the ability type in which 'to know what' and 'to know how' coincide.
3. Among the apparatus and technological theories referred to in point 1, the so-called intelligent machines and the corresponding information theories occupy an increasingly important position (cf. Zimmerli 1988, p. 357 *et passim*).
4. Accordingly life sciences have undergone a shift from analytical to what has been called 'synthetic biology' (Winnacker 1990).

We now have to turn to the present features of ethical reasoning as such. It is beyond any doubt that ethical arguments today are in great demand and that at the same time quite considerable changes have taken place not only outside but also inside of ethics! After a period of militant discussions between supporters and opponents of advanced genetic engineering methods in biotechnology, we have now reached a point that could be described as 'Farewell to spectacular lines of argumentation'. Neither the arguments of the technocratic Lederberg or Watson type (i.e., every action is permitted as long as it broadens our knowledge) nor the fundamentalist arguments of the Chargaff type (i.e., nothing is permitted because every action is an assault to the biosphere) have survived. A rather well-balanced and differenciated debate has taken their place; instead of all-or-nothing arguments, we are confronted with as-well-as or neither-nor considerations.

This allows for ethical reflections on a more sophisticated level, because even if a disagreement with respect to the content of ethical considerations should result, at least a fundamental agreement concerning the way and methods of deciding the respective question prevails. Correspondingly there is little or no dispute about the hierarchical order of the permissible types of ethical argumentation and about the possible realms of application within the field of genetics and biotechnology: Ethical assessment nowadays consists mainly of a reflection on the possible consequences of actions; therefore, the consequentialist (or teleological) type of ethical reasoning has superseded the deontological or metaphysical type almost completely. This does not imply that material values have totally lost their importance, but it does indicate their relative subordination to universally accepted procedural norms (see Sect. IV below). It goes without saying that in today's ethical discourse concerning the relation between genetics and biotechnology no unconditional praise or condemnation could thus be valid.

As far as the metaprinciples of a problem-oriented ethics are concerned, a model composed of formal norms and material values is the most promising. I suggest what I call an *ethics of situationally graded kinship responsibility.* In contrast to all anti-anthropomorphic rhetoric, I am of the opinion that an ethics of this type by necessity has to be anthropomorphic. Even the ethical considerations on the questions of genetics and biotechnology must be measured by the results of the ethical consideration regarding the application of advanced methods of molecular biology to human beings.

Due both to the intensification of science and technology and to the increasing complexity of almost all contexts, the applicational turn within ethics (DeMarco and Fox, 1986) is accompanied by a shift from principle orientation to problem orientation. As the problems under ethical consideration become scientifically and technologically more sophisticated (as in the case of research into the human genome), the analysis of the case itself is constantly gaining in importance. The fact that late modern societies have become increasingly pluralistic has also brought about a pluralization of values, which in turn has led to a situation where the traditional distinctions of theoretical ethics are of analytical use only. Even within academic discourse it is, for example, difficult to find purely deontological or consequentialist positions any more. As in a pluralist society different value systems and different approaches of justification coexist, the task of applied ethics to develop a procedural model of ethical consideration and decision-making becomes complicated to a degree academic theoreticans of ethics did never even think of.

III.

Some of my younger colleagues and I have therefore developed a four-step model of procedural ethics (Zimmerli and Wolf 1993, p. 316). The concept of a four-level model within a problem-oriented pluralist ethics (Zimmerli 1985, 1988 *et passim*) seems to be capable of meeting the conditions mentioned by ethically examining – step-by-step – different maxims, courses of action, and possible consequences without falling victim to the outworn dichotomy between deontological and teleological/consequentialist reasoning. The idea is to proceed top-down following the reduction in the degree of generality. Thus it becomes possible to avoid disputes on questions of principle where just questions of different possible applications are at stake. If an option has passed the examination of a higher level as being neither condemned nor recommended, it has to be scrutinized on the next lower level. Maxims, courses of action, or possible consequences that contradict the principles at a given level are to be condemned, those that completely comply with them are to be recommended.

Table 1. A four-step model of procedural ethics [from Zimmerli and Wolf (1993)]

Level/category	Forbidden	Yet to be decided	Required
(1) Formal principles			
(2) Temporally locally valid principles			
(3) Principles of professional ethics			
(4) Material values			

Level 1. On the uppermost level of generality, the main formal principles of modern ethics are to be applied: first the *principle of universalizability* (cf. Singer 1961), itself universally valid, substantiated deontologically by Kant and teleologically or consequentialistically by Bentham, Mill, and others; second, *the principle of equality* (or rather of equal treatment under substantially comparable conditions), and third, the *principle of justice,* formulated by Rawls as the *principle of fairness* (Rawls 1971).

A special problem is connected with a comprehensively formulated fourth principle, the *principle of humanity as humaneness.* By this principle, the fact is expressed that humanity is not something substantial except in relation to human beings. To examine whether some maxim, course of action, or possible effect is appropriate with respect to this principle, one has to take into account that human beings are not just spiritual or intellectual beings but very physical (and physiological) ones.

Level 2. On the next lower level we must integrate and then apply principles that are still generally valid, but in a temporally and locally relative way. Everybody knows that they are not valid all over the world and that some time ago they were not valid, as they probably will not be again in the future. But, nevertheless, everybody is convinced that it is necessary to have them now. Here, the defensive function of ethics, intended to impede the hasty and to slow down over-accelerated developments, comes to bear; consider for example the priority of the bad prognosis (Jonas 1979). This is a principle recommending worst-case scenarios and thus doing more justice to our situation than a negative ethics demanding a course of total abstinence.

Level 3. Next, the principles of the different professional ethics concerned are to be applied. These are – according to general consensus among the respective professionals – applicable to a certain well-defined area of cases. In the realm of the medical profession and of the application of recombinant DNA technology to humans, e.g., the "principle of informed consent" (requiring that no one may be subjected to medical treatment or scientific experimentation unless he or she has consented to it after being thoroughly informed) is professionally valid (cf. Nuremberg Code, Helsinki Declaration).

Level 4. On the fourth level, finally, the different material values of the relevant social groups, churches, parties, etc. are considered. Whereas on the first three levels a consensus is – at least in principle – attainable, on this level we have to start from the assumption that the examinations will as a rule end in *dissent.* Therefore, the procedural aspect is of uttermost importance, and this is where institutional solutions like ethic committees enter the stage.

IV.

The application of this procedural four-step model to the question of patenting human genes and living organisms results in the subsequent assessment:

(a) From a consequentialist point of view, no categorical objection could be raised against patenting procedures and products with respect to human genes, as long as no human germline cells are concerned. Some deontological objections, however, are raised by people who believe in special religious and/or ethical values, but these objections are either already solved (see Sect. I above) or at least do not prevent case-to-case evaluation.

(b) As far as the production of transgenic animals is concerned, again no categorical objection could be raised against patenting procedures, parts of or even whole animals from a consequentialist point of view as long as the positively assessed consequences outweigh the negatively assessed ones. Of course, in this case again some deontological objections exist (cf. OTA 1990, p. 130 ff.), as raised by different religious and ethnical groups. For the answer to these objections, see (a) above.

(c) The patenting of procedures and products with respect to genetically altered plants is even less criticized, although some ecological objections and questions as to the impact on the Third World have been raised from both the deontological and the consequentialist approaches.

What was argued comprehensively at the beginning with respect to fundamentalist criticism seems to prove true for the procedural examination as well. As far as the arguments are directed against *patenting,* they are unsound, and as far as they at least seem to have some plausibility, they are not directed against patenting but against the production of human genes and living organisms by the application of methods from molecular biology and gene technology. Most of them turn out to apply to questions of protecting disabled persons, of animal rights and of ecological ethics.

The metascientific objection regarding patenting as an obstacle to scientific research itself is obsolete; all notorious examples show that the contrary is true: patenting could even turn out to be a means of accelerating publication and making external application possible. Finally, patenting seems to provide an incentive of its own to research: many inventions came about precisely because of the intention to circumvent patents that already existed!

References

Barton JH (1991) Patentiertes Leben. Spektrum der Wissenschaft 5:74–81

Beier K-F, Haertel K, Schricker G (eds) (1991) Europäisches Patentübereinkommen, Münchner Gemeinschaftskommentar. Heymanns, Cologne

Böckle F (1988) Verantwortung für die Schöpfung. Chancen und Grenzen der Gentechnik. In: Die politische Meinung, Jg. 32, pp 45–56

Carter A (1989) The philosophical foundations of property rights. Harvester, New York

Cooper CC (1991) Making inventions patent. Technol Culture 32(1):837–845

DeMarco JP, Fox RM (ed) (1986) New Directions in Ethics. The Challenge of Applied Ethics. Rontledge, New York

Dolder F (1993) Patente für gentechnologisch erzeugte Tiere. Rechtliche Grundlagen eines Patentschutzes. Neue Zürcher Zeitung 19. 6., p 31

Domdey H (1991) Risiken der Anwendung der Gentechnik in der Tierproduktion: Biologische Sicherheit. Züchtungskunde 63(3):227–235

Einsele A (1993) Pharmaindustrie für umfassenden Patentschutz. Neue Zürcher Zeitung, 22. 6., p 13

Engel FW (1987) Sachschutz für Mikroorganismen nach dem neuen deutschen Patentrecht – zugleich Anmerkung zum BGH-Beschluß "Tollwutvirus". Gewerblicher Rechtsschutz und Urheberrecht 89(6):332–338

Freudling C (1993) Tierpatente in der Schweiz?. In: Goetschel AF (ed) Recht und Tierschutz: Hintergründe – Ansichten. Haupt, Bern, pp 161–199

Hahlbrock K (1989) Gentechnik in der Pflanzenzüchtung – Kann die Wissenschaft den Fortschritt noch verantworten? Max-Planck-Gesellschaft Jahrbuch pp 24–40

Hobom G (1986) Methoden und Ergebnisse der Gentechnologie. Bericht über die erste Arbeitsphase der synthetischen Biologie. Gießener Universitätsblätter XIX(1):17–27

Hohlfeld R, Nordhoff H-B (1988) Organismen als Produktivkraft. Ziele, Projekte und Probleme bio- und gentechnologischer Forschung in der DDR. Deutschland Archiv (Zeitschrift für Fragen der DDR und der Deutschlandpolitik) 21(2):182–196

Irrgang B (1990a) Hat die Natur ein Eigenrecht auf Existenz? Anmerkungen zur Umweltethik-Diskussion. Philosoph Jahrbuch 97 (2. Halbband):327–339

Irrgang B (1990b) Leitlinien einer Ethik der Gentechnik. Vorüberlegungen zu einer Ethik der Biotechnologie. Naturwissenschaften 77:569–577

Irrgang B (1991a) Sittliche Fragestellungen und Grenzziehungen bei der Anwendung der Biotechnologie in der Tierproduktion. Züchtungskunde 63(3) 247–257

Irrgang B (1991b) Sittliche Bewertungskriterien der Human-Genetik. Stimmen der Zeit 209(4):239–253

Jahnke L (1987) Patentinformation fördert Spitzenleistungen in Wissenschaft und Technik. Elektrie 41(3):83

Jonas H (1986) Prinzip Verantwortung – Zur Grundlegung einer Zukunftsethik. In: Meyers T, Miller S (eds) Zukunftsethik und Industriegesellschaft. München

Jonas H (1987) Ist erlaubt, was machbar ist? Bemerkungen zur neuen Schöpferrolle des Menschen. Universitas 42 1/2:103–105

Jonas H (1979) Das Prinzip Verantwortung. Insel Verlag, Frankfurt a.M.

Koechlin F (1993) Tierpatente sind unethisch. Neue Zürcher Zeitung, 22. 6., p 13

Koslowski P (1991) Genetisierung und Verlust der Gestalt. Folgen der Genetik für die Deutung des Menschen und der Gesamtwirklichkeit. Aus Politik und Zeitgeschichte (Beilage zur Wochenzeitung Das Parlament) 6:43–54

Lesser W (1987) Financing university research through patenting and licensing: Recent policies and practices at Cornell University as an example. Internat Rev Industrial Property Copyright Law 18/3:360–371

Löw R (1989) Die ethische Brisanz der Gentechnologie. Zeitschrift für Politik 36 (NF) (4):358

Long PO (1991) Invention, autorship, 'intellectual property', and the origin of patents: notes toward a conceptual history. Technol Culture 32(1):846–884

Mackley C (1987) The role of the patent system in technology transfer: the Japanese experience. Columbia J Transnat Law 26/1:131–166

Mahro G (1987) Rechtliche Regelung der Gentechnik in den USA. Eine kritische Darstellung des Rahmenentwurfs des 'Office of Science and Technology Policy'. Recht der Internationalen Wirtschaft 33(2):114–122

Marten H-G (1991) Gentechnologie zwischen Biologie und Politik. Interdisziplinarität und didaktische Struktur. Aus Politik und Zeitgeschichte (Beilage zur Wochenzeitung Das Parlament) 6:32–42

Merton RK (1957) Priorities in scientific discovery: a chapter in the sociology of science. Am Sociolog Rev 22:635–659

Moufang R (1989) Patentability of genetic inventions in animals. Internat Rev Industrial Property Copyright Law 20/6:823–846

Moufang R (1991) Ausnahmen von der Patentierbarkeit. In: Beier F-K, Haertel K, Schricker G (eds) Europäisches Patentübereinkommen, Münchner Gemeinschaftskommentar. Heymanns, Cologne, pp 1–91

Moufang R (1992) Protecting for plant breeding and plant varieties: a frontier of patent law. IIC Internat Rev Industrial Property Copyright Law vol 23:328–350

Moufang R (1993a) Methods of medical treatment under patent law. IIC Internat Rev Industrial Property Copyright Law vol 24:18–49

Moufang R (1993b) Patentierung menschlicher Gene, Zellen und Körperteile? Zur ethischen Dimension des Patentrechts. GRUR Int 439–450

Nevers P (1989) Forschungsziele des Kölner Max-Planck-Instituts für Züchtungsforschung. Auf dem Weg zu einer höheren Resistenz der Pflanzen. Max-Planck-Gesellschaft-Spiegel 6:37–49

OTA Office of Technology Assessment (1990) New developments in biotechnology: patenting life. Marcel Dekker, New York

Pechmann E Frh von (1987) Ausschöpfung des bestehenden Patentrechts für Erfindungen auf dem Gebiet der Pflanzen- und Tierzüchtung unter Berücksichtigung des Beschlusses des Bundesgerichtshofs – Tollwutvirus, in: GRUR Gewerblicher Rechtsschutz und Urheberrecht, 89. Jg., H 8:475–481

Rawls J (1971) A theory of justice. Harvard University Press, Cambridge

Redies B (1989) Freiheit und Risiken im heutigen Patentschutzsystem in der Europäischen Gemeinschaft. Recht der Internationalen Wirtschaft 35:958–964

Reiter J (1987a) Was kann der Mensch? Was darf der Mensch? Die Manipulation des Menschen durch die moderne Biologie. Caritas 88(4):164–173

Reiter J (1987b) Gentechnologie und Reproduktionstechnologie. Soziale Arbeit 36(7): 234–255

Reiter J (1990) Gentechnik und Patente. Soll Leben patentiert werden? Stimmen der Zeit 1:447–452

Ryle G (1949) The concept of mind. Penguin Books, Harmondsworth, London

Saedler RP (1986) Gentechnologie und Pflanzenzüchtung. Verbesserungen von Qualität und Quantität. Futuristische Bilder bald Wirklichkeit? Max-Planck-Gesellschafts-Spiegel 4:8–10

Säger M (1991) Ethische Aspekte des Patentwesens. GRUR Gewerblicher Rechtsschutz und Urheberrecht 4:267–273

Schenke G (1990) Gentechnik und Homo sapiens. Biologie in der Schule 39(6):209–216

Schweikhardt F (1986) Zur Rolle der Patentinformation an deutschen Universitäten und Hochschulen. Eine Umfrage des Komitees FID/PD. Nachrichten für Dokumentation 37(4–5):226–228

Singer MG (1961) Generalization in ethics. A. A. Knopf, New York

Straus J (1991) Der erfolgreichen Erkenntnissuche soll die Verwertung folgen. Für mehr Engagement zur Patentierung vor Forschungsergebnissen. Garching Instrumente als Partner des Grundlagenforschers. Max-Planck-Gesellschafts-Spiegel 1:12–15

Sullivan RJ (1989) England's 'age of invention': the accleration of patents and patentable invention during the industrial revolution. Explorations in Economic History 26/4: 424–452

Vitzthum Graf W (1987) Gentechnologie und Menschenwürdeargument. Zeitschrift für Rechtspolitik 20(2):33–37

Vossius V (1990) Patentschutz für Tiere; 'Krebsmaus/HARVARD'. GRUR Gewerblicher Rechtsschutz und Urheberrecht 92(5):333–337

Wassermann U (1986) Patents and the biotechnical industry. J World Trade Law 20/6: 705–713

Winnacker E-L (1990) Synthetische Biologie. In: Herbig J, Hohlfeld R (eds) Die zweite Schöpfung. Hanser, Munich, pp 369–385

Zimmerli WC (1986) Prinzipien einer nicht Prinzipien-orientierten Ethik. In: Arbeitsplätter für ethische Forschung, 15. Jg 1, pp 2–16

Zimmerli WC (1988) Krise der Krisenethiken. Moralphilosophische Engpässe im technologischen Zeitalter und das Konzept einer problemorientierten Ethik. In: Kluxen W (Hg) Tradition und Innovation. XIII. Deutscher Kongreß für Philosophie (Bonn 24.–29. 9. 1984), Hamburg, pp 355–370

Zimmerli WC (1992) Privateigentum – etwas eigentümlich Menschliches? Eine ideengeschichtliche und philosophische Analyse. In: Dichmann W, Fels G (eds) Gesellschaftliche und ökonomische Funktionen des Privateigentums. Deutscher Instituts-Verlag, Cologne, pp 263–292

Zimmerli WC (1993) Von den Pflichten möglicher Eltern und den Rechten möglicher Kinder. Ethische Dimensionen des Heterozygoten-Trägerscreenings. In: Toellner R (ed) Medizin-Ethik, vol 4. Fischer, Stuttgart

Zimmerli WC, Wolf S (1993) Die Bedeutung der empirischen Wissenschaften und der Technologie für die Ethik. In: Hertz A, Korff W, Rendtorff T, Ringeling H (eds) Handbuch der christlichen Ethik, vol 1. Herder, Freiburg, pp 297–316

Discussion: Session 4

Straus: I think this was an excellent presentation and would like to hear Dr. Grunwald's comments.

Grunwald: I think it's quite important to see that from Dr. Zimmerli's ethical point of view there's no objection. I agree and think it's not primarily a question of good or bad. The formal principles you showed us are exactly the principles that were discussed yesterday. One of those principles was to balance interests and this is what you say, to respect the principle of the worst prognosis, for example. I couldn't agree more: Let's say this is not only a question of good or bad, it's more a weighing of facts and interests.

Beyreuther: You didn't touch germ line mutations and germ line manipulation. I think it's correct if you don't touch it, but I like to touch it. There's one very interesting point that we have to consider. It's the difference between females and males. Males more or less don't age in regard to reproduction. Females do. If they have used up their 1000 eggs they're old. Now why do we not discuss whether science is needed in the context of the topic covered here to come with some solution of this problem. Don't females also not want to age? Isn't it something that we have to discuss now?

Chairman: Of course we could discuss many different questions. I do not quite see how it interrelates to patenting.

Beyreuther: I think that this is something regarding the genome and genes and its introduction. And I think it is important to discuss whether those genes that regulate the developement and regulation of the reproductive system should be taken out or not.

Zimmerli: Well, of course, but you wouldn't do that by forbidding patenting. You would forbid application, so this wouldn't be a question of patenting, it would be a question of the gene law or embryo law.

Beier: As a director of an institute for and not against patent law, I am very satisfied to hear from you that you don't find anything unethical in patenting, but the legal consequences of that standpoint would be that we have also to cross out those provisions in our patent laws, European patent, laws, national patent laws, which forbid the patenting of inventions the exploitation or publication of which is against public order or morality. Only for a better understanding of these provisions I would go back a little in history. These provisions have been introduced in the last century when most of our patent laws have been drafted and this provision had some Victorian approach for a long time. You know, the patent offices would not publish, for instance, a contraceptive, which was against public order even if the patent law always gave no permission to use that, but to forbid the use of it. And please remember the arguments of Greenpeace a.s.o. – that if the state, if the government gives some title to person, then there is a positive approach to that invention involved. And we cannot always repeat: "No, no, it's not up to the patent system to control undesirable social consequences of inventions with the stamp of the state and with the granting of the privilege it is always the idea that that invention has been approved by the state. Historically this was the main effect of inventors's privilege in the Middle Ages. The inventors' privilege in the Middle Ages gave the owner not only the power to exclude others from using the invention, but it gave the positive right to use the invention against guilds a.s.o. and I think this attitude is still in the heads of many people. Therefore, we can always repeat and we must repeat that this is not the function of the patent system but we don't find believers and in a hearing I spoke with an assistant from the University of Bonn who was sent by the greens to that hearing and I explained to him this situation. His answer was: "I know very much that the patent system itself has not the function to control uses, but if I have read your papers correctly, the patent system is for fostering innovation, for promoting research and development, and if we include biological inventions in that system, we will promote research and development in these fields and this is just what we don't want." This was a very clear answer, but it is an answer which put the problem from the patent system on another level, namely on the level of research and development.

Zimmerli: I agree completely with the second part of your statement. I do not think that the first part of your statement is conclusive. I think there might be reasons to eliminate the restrictions out of the European patent regulations, but not because of my argumentation because, as I would like to point out, there is no logical relation of deduction between ethical principles and legal qualification, just the relation of non-contradiction. So it shouldn't contradict, but if you include some kind of restriction into a law, that doesn't restrict the fact that there are no ethical objections. It doesn't follow logically. So this brings me to a second remark. Of course our legal system implies or goes back to many different sources. It is an act of legislation and therefore, it is an act within our political system and politicians do not only consider ethical questions. They do consider different questions, for instance, that there are quite a lot of fundamentalists around, which is O.K., I mean we're living in a pluralist society, so why don't give fundamentalists the right to live there as long as they don't prevent other people from doing their work. And they are voters, of course, so if you want to have a law which guarantees your legal system to survive, then of course you will take into account that there are some fundamentalists around. And therefore, we have very different types of legislation with relation to nations. If you look, for instance, at France –

very surprising in France – if you look at, on the one hand, the attitude towards new technologies, especially nuclear technology, and then you look at their attitude towards life sciences at the other hand. It's quite the contrary. But that's how France is, so we have to really look at this in this four level model, look at the different regional principles from, for instance, the point of view of the national systems of legislation.

Gugerell: I would like to come back to the proposal from Prof. Beyreuther to discuss changes in the germ line and you said that there is no patenting. The position is this, in Germany there ist no research allowed and therefore, no patenting will arise from here. But my point is that somewhere in the world it may be possible to do it and interest may then be to also patent it in Europe, for example in our office. So I suspect that sooner or later we will get cases on gene therapy affecting the germ line. We do already have a great number of applications for somatic gene therapy. And therefore, it is of interest.

Zimmerli: I, of course, would have loved to comment on this. Just a quarter of a year ago we had this conference in Erlangen on somatic gene therapy questions and again I'm of the opinion that there are some ethical considerations necessary to deal with the question which applications of somatic gene therapy, for instance, are ethically sound and which aren't. But that would in my opinion not afflict the question of patenting them because, of course, the patent law forbids patenting things which are against morality or public order. So, we have already a tool of discriminating between those patents which might have consequences which are undesirable from a social point of view and those patents which do not have those consequences. Another question is – and this, of course, applies especially to germ line therapy and therefore, I would still rule germ line therapy out of the possible applications – whether we can perceive the consequences, and of course with germ line therapy, we can't. We only can if we do it. So, we have to commit the "crime" first in order to be afterwards in the situation to maybe legitimize it. This, of course, is the wrong way around. And therefore, we should completely rule out the application to germ line therapy and we should with respect to somatic therapy (and, as I would like to add, diagnostic application of genetic tools) make a very careful case to case evaluation. There's no other way. We don't have one rule that solves all the problems. So, I would be very interested in discussing the question of somatic and germ line therapy, but as far as I can see, it does not afflict the answer I gave with respect to the patenting question.

Straus: I was very pleased listening to you in most respects and one of your remarks was particularly valuable. Our colleague, Grunwald, yesterday made a few remarks about possible obstacles or hindrances due to patents in scientific research and you mentioned that patents do quite the opposite. I would add that we are not in the 19th century and the problem is there whether we have a patent or not because of the proximity of the research to the possible commercial application. And as soon as that is there you will not get your material without having signed some agreement or offering more or less the same valuable material in exchange for that.

Zimmerli: I would agree although I would say that this is probably not a consideration within ethics but a rule of prudence. I just want to add, if I may, a short comment to your paper which you sent me, Dr. Vogel. Of course, because of the fourth level I mentioned there are very sound positions which raise objections – which is O.K. But these are, as I would call it terminologically, moral objections, not ethical because the difference between moral and ethical is that ethics is that little bit, as my teacher Günther Patzig, would call it, that little bit of morality we can grasp by rationality and there are lots and lots of not irrational but nonrational motivations included which are not capable of being grasped rationally. So, I agree with some of your points without introducing them to my own reasoning because they wouldn't follow from my principles.

Hofschneider: At this point of the discussion I would like to address a question to Dr. Vogel. You organized the meeting and you are very grateful for this in particular since you have ethical concerns with respect to patenting of human genes and living organisms. Now how do you feel after having heard that contribution?

Grunwald: As Dr. Straus addressed me I'll try to repeat what I said yesterday because Dr. Zimmerli wasn't here. I'm administratively responsible for the Deutsches Krebsforschungszentrum. I said yesterday we see some problems with a growing discussion and the public uneasiness about patenting and do not as clearly as our learned colleagues agree with the concept of patentability of, for example, animals; let's therefore concentrate the discussion on the onco-mouse. We clearly foresee that it's going to take years before we have a clear-cut legal situation and the outcome is not completely clear. It will take about two years in the European Patent Office and Judge Bruchhausen will sit in that committee which has to discuss the case. Probably it will go to the legal courts, so we'll have litigation. Our colleague from the EPO said he sees at least three, perhaps four years of uneasiness, so my contribution yesterday said we see problems and issues raised for science, especially basic research. We have an interaction within the worldwide science community which up to now was free. We did not even have declarations that we accepted from firms that what they sent us only belonged to them, but that what we do with it belonged to them, too. This is changing. We get such material transfer agreements from universities now, not only from industry. We see that concerning organization we have to keep materials under lock. We have bookkeeping of materials and we have the problem that we have to report on things we otherwise wouldn't report. For the types of science concerned, it's much more project and result oriented than good for this science. The search for truth which was, at least in life sciences, up to now a very basic thing will, I think, not be abolished. I don't take patents that important, but it'll be a difficult situation. So, this was just to repeat what I said yesterday and so on this level I don't see any ethical questions: I don't think bad and good are criteria here. But what I wanted to say is, the general clauses are the forming stages of law. You said correctly, "what's not yet the law" and the discussion has to be, do we find things to decide under these general clauses. The onco-mouse case was decided on that general clause of morality. That says, we might need legislation. There I would be interested in your position. I believe in piecemeal social engineering and I think that we have to see what you suggest for the voting people. We heard yesterday the colleague from the European Commission. He suggested piecemeal social engineering and addressed politics as the art of the possible. Our problem in Germany, at least, is that some experts think that they know everything and that they have the solution and everybody should accept it. To put it in a nutshell, they see the human genome as the sum of chemical products to be patented. And they can't sell that to the public. So, what would you suggest?

Zimmerli: The first question I would like to address as an ethicist. The second question I would like to address as a philosopher in an non-ethical respect. If I recollect correctly what you said, then you said that there might be some bad – I formulate it from an ethical point of view – consequences, not evil, but bad consequences for the scientific community if too much patenting intrudes upon what we call basic research. Now the question ist whether we take into account correctly what's going on within science. We are just putting through research for the National Ministry of Research and Technology with respect to what we call the change of the type of knowledge within the life sciences – we surveyed different institutions from basic research to industry and it's surprising that if we look at what people actually do, not what they are saying, most of them are applying methods and developing methods and are doing what one would think only people in the technological field do. Very few people are therorizing, very few people are competing for nobel prizes, many are competing, but very few do what we usually connect with the breakthroughs in the basic sciences. And this, of course, is what Popper hat to learn from Thomas Kuhn: that most people do not do what he thinks scientists do. Most people are just applying methods, nothing else. Now, if this is the case and this is especially true within life science because, as you all better know than I do, there has taken place a paradigmatic shift from the, as it were, outstanding paradigmatic science called physics to the outstanding paradigmatic sciences called life sciences. An with this paradigmatic shift, also the shift from a basic research oriented type of science to a more or less application oriented, highly technologized research takes place. If you think, for instance, of Venter's new institution that he has about five or six times more robots than human beings doing the sequencing and the computer enters life sciences to a degree never heard of before than it becomes evident that we actually have quite a new type of science, our life scientists becoming informatic specialists and informatics is a highly technologized science which is very close to technlogy and not so close any more to mathematics. So, what I am saying is that we probably will not have to protect basic research in the same sense than we used to. So, our protection of basic research in the future will be, as I think, more of the type of protection by patenting. Now I'm arguing positively whereas I argued negatively during my talk. It would be responsible if people would look at patenting as a way of improving our possibility of this new type of life sciences. The second question, what would I say if people would ask me, and this of course connects with my first answer, what the situation with respect to legislation is. I would first say, we are not suffering from a lack of legislation, we are suffering from over-legislation. Our system is an over-legislation system and doesn't give enough room for individual decisions. So, I would say: as much legislation as necessary and as little as possible. Therefore, I am not quite sure whether we should introduce more legislation into our scientific system. I think we should introduce more ethics in the sense of self-reflection and autonomy, self-determination. So, I strongly insist on improving the professional codes of behavior. I think this is much more important than to alter legislation because if each time when science is developing further we alter our legal system then, of course, it always comes too late. But if we do it by ourselves, if the scientific community follows the rules of self obligation to do some things and to not do other things that would be much more effective and ethically much sounder. Just wait for the legal system to react. So, that would probably be what I would recommend – that the scientific community of life scientists or of molecular biologists or however small you want to cut this scientific community would again try to develop some kind of code of behavior with self-obligations.

Vogel: Dr. Hofschneider has asked me an urgent question. I agree with this reasonig in four levels (Zimmerli). Let us start with the first formal principle. The principle of equality is very important. Not everybody has to be equal, but everybody has to be treated equally and in close connection with it the principle of justice and fairness. And what we have heard so far, especially in plant breeding, the principle of patenting in plant breeding, you remember the discussion on farmer's rights a.s.o. and on breeders' rights – many of the things which are being done in this field are violating severely the principle of justice and fairness. Once we stay within the level of the first world countries, this might not be so obvious. But as soon as we introduce the viewpoints of the developing countries, then the principles of fairness and of justice are violated very severely. I think this is one point I wish to make. And then the principle of preference for the worst case scenario. I think this is what I had in mind. In many things I had to say I have learned from Hans Jonas and I think all of us have to learn from Hans Jonas one way or the other. I tried to think out the worst scenarios in this field of patenting human genes and this has been my reason for opposing patenting of human genes. My worst scenario is not so very unlikely: If the genes as such are patented, the genes can be used not only for the very harmless and benevolent use for making functional proteins of whatever, but can be used for an unknown number of purposes we cannot know so far and it is the problem of all prophecy that we can only extrapolate from what we know now. However, the unexpected things are happening and you look at the progress of molecular biology – so many unexpected things have occured during the time I worked in the field of human genetics. Therefore, I expect that many more unexpected developments will occur and in situations like this ownership of a patent for a human gene or for many human genes – and the numbers you have told us from Venter's study and yesterday we hat the Adler paper, he also gave the same numbers, – it is really very, very, difficult to think about the consequences of all this and to think about all these scenarios that are possible. Therefore, I think that very careful procedure in patenting is very important. The traditional aspects of patenting laws do give us such possibilities. Dr. Gugerell tried to convince us yesterday in a long discussion that human genes can be invented or can be parts of an invention and I said they can only be discovered and therefore, they cannot be patented because I have them in all the cells of my body. This is not so much theoretical obstinacy – when I said this it was a kind of idea how to draw practical consequences, using the obviously good consequences of patenting, but to avoid the worst case scenarios of the future. The way I suggested is to patent human genes or gene sequences only limited to certain procedures, for example, producing inteferon which might be good for cancer. When you actually remember your own principles and talk to people, even the principle of the worst scenario would help you to be very, very, reluctant towards unlimited patenting. The same holds true for patenting of cDNA sequences which are completely unknown. Let me repeat, I'm not at all against patenting, I am against unlimited patenting, I am against patenting in fields where the first principles of justice and fairness are violated and secondly where the principle of the worst case scenario is violated. A second general principle – patenting is not only a matter between scientists and commercial users, it's a matter of the society at large and it has to take into account the psychology of society at large. My worst case scenario on this point is that we are living in a society where an increasing number of people have the general idea that scientific research as such and life science, are a bad thing and have to be limited. I shall conclude with a metaphor: When I'm forced as a scientist to go across a meadow where there is a wild bull, I certainly will not go with a red coat.

Sela: I would like to second Prof. Vogel on the patenting of human genes and I distinctly comment on human genes and distinguish between human genes and other genes and I would like to address it from the point of view – I'm not sure if this is philosophical or ethical. If anybody patents a human gene per se he claims, at least for a limited period, to be the proprietor of this gene. As a matter of fact, he's claiming something which is already my property. This ist form the ethical point of view. I would like to accentuate it by an absurd example. Even if he claims a utility patent on that gene, let's assume an human mutation and I am making in my body a tremendous amount of insulin and somebody has the patent of the human gene for insulin, even an utility patent for the human gene for insulin, is the insulin that I make in my body my property or his?

Lange: I just have a question of understanding. In your procedure model, the principle of respect of nature, what does it mean and which consequences has the regional principle in our context here?

Hess: I think after all we in our own bodies have billions of E. coli. If we extinguished this E. coli we would be dead within 24 hours. Now, do we have any respect for E. coli because after all it's our nearest neighbor? Perhaps this goes along with the question you asked, but I have another point of view with respect to these categories. It would be useful, at least for me, if you could just say a few words about which of those principles are indeed local principles or regional and which are general in all cultures of this world because I encountered all over again especially with respect to some of the points of view which you mentioned in the second category that something which is accepted and even legally enforced, let's say, in Sweden, is absolutely not acceptable in Japan. Now these differences certainly are extremely important for evaluating any type of categorial analysis along these lines because the weighting of those different levels is a little bit tricky. Obviously it does not hold true for the whole area of patent and this is a very interesting development because the patent law is internationalized or it is on its way and certain parts of the culture and the other part which involve legislation and weighting of local ethics a.s.o. are absolutely decentralized and not homogenous.

Hofschneider: Now I will try to formulate my own question briefly. You said the basic question is, is patenting ethical or not ethical. I think you came to the conclusion that it's ethical and one important reason was that a patent provides useful information. From my point of view, and I would say from the point of view of many scientists, patents do not provide useful information to reproduce the results with reasonable effort. Now if that is frequently true someone has again to ask whether the ethical basis for patenting is given at all.

Zimmerli: First, Dr. Vogel, I completely agree with almost all the points you made, but I completely disagree that we should handle these points by restricting patenting. Of course, we should do something against the inequality between our world and the third world, but we can't do that by patenting. Of course, we should have a worst case scenario and of course, there are horrible uses imaginable, but we can't fight them by the patent law. I differ in one point, the differentiation between invention and discovery being already settled in the patent law. It's a question of how we interpret it, but it's not a question of new legislation, because we have this legislation already. So, this is again no question of ethics with respect to the law. So, I would say we could probably agree point by point and we could probably even agree whether we should solve this by patent law or not and I think at the end you would agree with me that this is not a question of patent law. Rather a question of society at large – you are completely right. Society at large is showing a certain behavior of acceptance and of non-acceptance and I'm the first to agree that technology assessment should include social acceptance and of course, ecological viability. But again this is not possible to rule by patent law. You cannot change the acceptance or non-acceptance by changing the patent law. But what you have to introduce positively and what most technology assessment institutions already do is risk discourse. You have to institutionalize a risk discourse between not only the scientists and not only the politicians but also the other groups of society concerned. I agree with that. So, because we disagree we agree. Dr. Sela, it's not the DNA sequences that are yours that are patented. You still are the proprietor of your DNA sequences. But what is patented is a certain procedure or a certain kind of artificial product which you don't have. So, this doesn't quite apply. Of course, you have insulin, but still we would allow somebody to patent a certain procedure to produce insulin. I see no contradiction at all. Maybe I should comment on Dr. Hess first. First of all, let me answer the joke he made at the beginning. I didn't know that you are such a close relative to Escherichia coli, but I hear that with interest. I was talking about biological relationship. The main point I think was the regional differences and this is a very important point, actually this whole procedural model is only necessary because we have regional differences. If we wouldn't have them it wouldn't be necessary, we could just apply the principles and therefore, I think what we should answer is the question Dr. Lange and you both asked correctly – the principle of respect for nature. This principle is a principle you find in different cultures at different times. Therefore, it's a regionally and temporally limited principle. It hasn't been valid in all cultures at all times, it hasn't been in our culture, for instance, for a long time. But now it has become valid together with what Hans Jonas asked for – and Hans Jonas, as you might know, was during the last 10 years a very close friend of mine – and I think he was completely right in asking for the worst case scenarios. He was completely right but he wouldn't have been right a hundred years ago because a hundred years ago it was the wrong thing to do to go by worst case scenarios. Nowadays we have to go by worst case scenarios because we have already developed some of the advantages of technology and now we have to avoid the disadvantages. But you can't avoid the disadvantages before you have the advantages. So, that's what I wanted to say and now I come to the last question which I, frankly speaking, didn't quite understand because, to my knowledge, patents do not distribute this information. Otherwise I couldn't see why so many patents are quite nicely used in industry and technology.

Hofschneider: I can give you an example, that is the patenting of hepatitis C virus. If you order the patent for a few hundred dollars you do not get much information. The main part of the information which you need to perform experimentals successfully is not in the patent in spite of the fact that the patent owner has the knowledge. If you request the clones, you don't get them. If that is a very frequent thing, then I would think you would have to consider whether patenting still can be considered to be ethical. It know it's a disadvantage of ethical rules that they are usually not followed. Anyway, one should discuss it.

Zimmerli: I understand what you mean but I would put it differently. I would say, the ethical question included here is the question whether the patenting was correctly done. So, it's not an objection against patenting, it's an objection against not correctly doing the patenting.

Straus: I was suspicious that this was also the problem of your patent and of course, we could name a number of such cases. We tried, that means Dr. Moufang and I, just recently to convince the European Patent Office that in that particular case the description was not sufficient. But in any case this is not the failure of the system but it's the failure of a particular applicant and the patent office should not grant patents, if you are right, and in addition you should try to get the clone from the ATCC if it is deposited and if not, try to convince our General Administration to start a nullity case to revoke the patent. This would be very costly but maybe successful.

Session 5

The Patenting of Genes and Living Organisms: The South's View

Suman Sahai

I would like to introduce into this discussion an eco – political aspect of patenting genes and living organisms. I refer here to the viewpoint of the developing countries with particular reference to the use of plant genetic resources. Plant genetic resources have always been a valuable economic resource but with the advent of biotechnolgoy, they have become immeasurably more important. They are thus the target of more and more stringent instruments of privatization, whether through tightened Plant Breeders Rights (PBRs) or now, through patents.

Ownership of Genetic Resources

While the rush is on to patent every conveivable gene and organism from the plant world, it would be worthwhile reflecting on the ownership of these plants and genes that are the subjects of patents, mostly by multinational concerns based in industrialised countries. The genetic resources of this world are a natural resource on which several forms of economic activity like agriculture, pharmaceuticals, textiles and other forms of industry are based. In this they are no different from other natural resources like copper, coal or oil.

The international convention governing the ownership of natural resources acknowledges that natural resources are the property of those countries in whose territories they are found. Thus copper found in Canada is the property of Canada and the coal found in Germany is considered German.

This international convention is curiously ignored in one solitary case: that concerning the ownership of genetic resources. These resources, also referred to as biodiversity, are located almost exclusively in the tropical and subtropical areas, that is in today's Third World. The great Russian scientist Vavilov who studied and documented the origin of almost all known crop plants has identified the centres of genetic diversity. Except for Australia, all the other centres of diversity are located in the developing countries (1).

The North is a truly gene poor area, and its dependence on Third World germplasm cannot be exaggerated. For example, every Canadian wheat variety today contains genes introduced from as many as fourteen different Third World countries. American cucumbers use disease resistant genes from Korea, Burma and India. New Spinach varieties are protected by genes from India, China, Iran and Turkey (2, 3, 4).

> "The importance of biological diversity to the future of genetic engineering can not be overemphasised. Germplasm is the fundamental resource of the new biotechnologies, and it is conceivable that it could become a limiting resource. . ." Mr. Murray advising the US to negotiate with the Third World for access to genetic resources added, "However, many developing countries will have an advantage in that a large percentage of the germplasm resources of the world are located in developing countries. (James Murray, Strategy Conference on Biological Diversity, USA 1981)

> "If we had only to rely on the genetic resources now available in the US for the genes and recombinants needed to minimise genetic vulnerability of all crops into the future, we would soon experience losses equal to or greater than those caused by Southern Corn Leaf Blight several years ago . . . at a rapidly accelerating rate across the entire crop spectrum" (J P Kendrick Jr. Univ. of California, Davis, USA)

Genetic Resources however, unlike all other natural resources have been declared to be the Common Heritage of Mankind (5). This has led to the blatantly unfair situation in which use of these resources has never been paid for. So whereas Canada and Germany were able to earn from their copper and coal, the Third World was denied the opportunity to earn from its biological diversity. If the copper found in Canada is Canadian and the coal found in Germany, German, then by the same principle, genetic resources found in the Third World will have to be considered their property. There cannot be two sets of rules.

Nor is it the location principle alone that makes genetic resources the property of the South. These resources have acquired their recognisable, economically important form because of the labour and ingenuity of Third World farmers and indigenous people. There were no plants of rice and potato and cotton lying around in the forests waiting to be picked up. Food and cash crops on which the survival of the human race is based, were created from wild plants in tropical forests by generations of careful breeding and selection. This innovative, laborious process carried out by the farmers of the Third World has gifted the world a stable, secure food supply and many grain dollars to countries exporting agricultural surplus.

It is not only that Third World farmers have created almost all important crops, they have also identified in several cases, the genetic traits that gave these crops desirable characteristics. So it is that the first disease resistant potatoes were bred by the farmers of Bolivia. To the West African farmers goes the credit of breeding some of the first insect resistant bean varieties. Rice and certain forms of wheat were created by the farmers of India. Ethiopia, the home of coffee has also given us some of the earliest resistant strains of barley (6, 7, 8). Yet these enormous achievements by generations of Third World people have never been acknowledged, while the fruits of their labour have been approprited without compensation.

Having helped themselves to these resources for free, the countries of the North have proceeded to earn substantial sums from them. Germplasm import to the US is reported to have fetched the American economy upwards of 70 billion dollars from agriculture alone (9). The millions being earned from plant based drugs taken from tropical regions are another chapter (annual turnover: 60 billion and growing) as are all the varied industries based on living resources (10).

It should be obvious that the principle of Common Heritage cannot continue to be supported, more so in the light of the growing trend of privatisation by patents and stricter Plant Breeders Rights (PBRs). If genetic resources are a common heritage, they cannot be privatised. If they are to be privatised, they must be acknowledged as the property of the Third World and paid for like any other privately owned resource. A correct step in this direction has beenn taken in the Biodiversity Convention which was signed by most nations at the Earth Summit in June 1992. This treaty acknowledges the sovereignity of nations over the biodiversity found in their territories. This principle will have to be extended to the arena of patents on genes and life forms to help define guidelines for an Intellectual Property Rights (IPR) system that is just and serves the interests of all parties.

Return on Investment

Corporate interests defending their demand to patend put forth the argument that strong forms of protection are necessary to secure returns on the large investments they make when developing biotechnological products by using genetic material. This is a valid point. However it cannot be any one's case that corporations should be allowed to appropriate raw materials for their productions, without due payment. That ist what they do if they do not pay for genes that serve as the foundation material on which their large investments are based, and from which they expect large returns.

Consider the following facts: A few years ago the musk melon crop of California, valued at millions of dollars, was threatened by a fungal outbreak. Resistant genes were brought in from India and the Californian musk melon industry was saved. India received no payment for the genes from which the Californians earned several millions. The rosy periwinkle has become a cliche in the lore surrounding exploitation of diversity. This plant native to Madagascar was found to contain two drugs called vinblastine and vincristine, still the only known drugs effective against a kind of childhood leukemia. The pharmaceutical giant E. Merck patented the drugs and destroyed the natural habitats of the periwinkle by overexploitation. Madagascar never received a penny (11, 12).

Or the case of resistant genes against the viral Yellow Dwarf disease in Barley and Wheat brought in from Ethiopia to Europe and the US. WO Grace, an American company has patented the active principal of the "Neem" tree, to use as a biopesticide although Neem has been used as a disinfectant and pesticide since centuries in India (13, 14).

Many would be surprised to know that the United States in the second largest exporter of rice in the world after Thailand. All the rice germplasm is taken from the Indian and Japanese centres. No payment of any kind has ever been made for the use of germplasm from which the US has profited immensely (15). The list is endless. I have only used some illustrative examples to make the point.

I do not believe that honorable people can find it possible to support the thesis that this highly objectionable plunder of the wealth of developing countries should continue unchecked. And what is more, that this plunder should be institutionalised on the global stage by forcing patents and PBRs through the threat of sanctions and penalties in trade fora like GATT.

Technological Development and IPR

The level of intellectual property protection that each country allows is directly related to its level of development. The more technologically and economically developed a country, the stronger the level of intellectual property protection it accomodates. In Europe we see a system of IPRs for plants progressing from PBRs which were at first flexible, with implied or outright exceptions for farmers and plant breeders, becoming stricter by curtailing the farmers and breeders exemptions and finally introducing the highly monopolistic form of protection by patents.

Under the older convention of UPOV (Union for the Protection of New Plant Varieties) which was constituted in 1961 and revised in 1978, a new plant variety produced by a breeder could only be produced and marketed by him. The breeder had a monopoly over his variety via the seed trade but the system allowed two important exemptions. One was called the Breeders exemption, which allowed other plant breeders to use the protected variety for breeding purposes; the other exemption implicit in the UPOV treaty was that for farmers. The Farmers exemption gave farmers the right to use seed from their harvests to plant the next crop, even if the seed was protected by PBRs (16).

In 1991, the UPOV treaty was revised once more and the Breeders exemption was almost done away with. This means that now breeders have to pay royalty to the PBR holder if their new variety bears some resemblence to the protected variety even if the new variety has been bred for different traits. The farmer after the 1991 amendment cannot use farm saved seed from protected varieties, without some compensation (17). The methods of compensation are being currently discussed in various fora in Europe and the subject is a bone of contention between farmers and breeders.

With the advent of recombinant DNA technology, a new kind of plant breeder has entered the field of plant breeding. This is the biotechnologist who uses gene technology to introduce genes into plants by modern molecular biological methods, to create new varieties which are described as "transgenic plants". Although the new kind of breeder performs essentially the same task as the conventional breeder (that of shifting genes around to create favourable combinations), he has more sophisticated tools. His greatest advantage however is the ability to shift genes across the species barrier.

This means that whereas a conventional breeder cannot cross a tomato plant with a corn plant in search of a favourable gene, the biotech breeder can cut out a disease resistant gene from corn and put it into tomato to confer resistance and make a transgenic plant. This category of breeder has demanded and received protection of the gene and the new variety created by its transfer, through the very rigid form of industrial patents. This in my view is totally incorrect since the breeder producing transgenics can at best ask for a process patent but not a product patent for the gene which is not his invention. Industrial product patents are the strongest, most rigid form of protection that exist. We have thus seen the progression in Europe from 1961 and lax PBRs to the 1990s and the introduction of industrial patents for plants and animals, in keeping with increasing strengths in biotechnology.

The situation is similar in the chemical and pharmaceutical sector. When the technological development in industrialised countries was not as strong as it is today, only process patents were applied. This meant that if someone invented a new product like a new drug, the person could patent the process by which the drug could be produced, but not the product itself, that is the drug. There were good reason for this. Allowing only a process patent enabled competitors to find better, improved processes for producing the same or similar drugs, perhaps cheaper or with fewer side effects. Having only process patents thus allowed technological development and upgradation of product quality. Technological improvement would not be possible if the product could be patented and a monopoly established on producing the drug itself. Once the technological strengths increased and became established, industrialised countries began to introduce product patents. This made is possible for them to market their products in other countries under monopolistic conditions, giving the host countries no chance at all to compete with the same or similar products of their own in their own markets.

Industrialised countries kept for themselves the choice of when to switch from process to product patents, making the switch only when they felt they had acquired competitive advantage. In fact a G7 nation like Italy introduced product patents as late as 1982 and Spain even later, in 1987, to allow their chemical and pharmaceutical industries to become technologically competitive before introducing product patents.

Which principle of fair trade would allow one set of conditions to prevail for the industrialised countries and quite another set for the developing countries? And yet that ist what is being sought through the efforts to "harmonise" Intellectual Property regimes across the world regardless of the respective stages of technological development in individual countries.

By demanding that the same rules for Intellectual Property protection apply in India, Bolivia and Egypt as in Japan, USA and Europe, the industrialised countries become vulnerable to the charge of Economic Imperalism. Surely if the

Newly Industrialising Countries (NICs) of yesterday like the USA, had maintained lax IPR systems as protection from the technological domination of the then technology giant England, then they should not be averse to letting the same principle apply now to the technology dependent countries of the developing world.

What is of some concern to the people of developing countries is also the fact that the protagonists of patenting in biotechnology are to a large extent huge transnational corporations, in particular the agrochemical giants, who have acquired a big stake in the commercial exploitation of genetic resources (see Table 1).

Table 1. The Chemical Seedsmen

Company	World Rank In		
	Pesticides	Pharmace-Uticals (1987)	
Bayer (D)	1	2	Active in plant breeding and seed production particularly for herbicide and pesticide resistance
Ciba-Geigy (CH)	2	4	
Shell (NL/GB)	3	–	
Monsanto (US)	4	–	
Eli Lilly (US)	8	11	
Du Pont (US)	9	–	
Stauffer (US)	10	–	
Hoechst (D)	11	1	
DOW (US)	12	33	
Sandoz (CH)	19	7	

It is known that these multinational companies have enormous financial stamina and very powerful marketing facilities. Their entry into any economic sector poses highly unequal competition for other, smaller players. If the harmonisation of IPR regimes is enforced, seed producers in developing countries, who are still to a large extent farmers, would be required to compete with multinational companies with huge financial resources under conditions and rules like the stringent IPR systems that have been developed in the industrialised countries to protect their corporate interests. This does not constitute a level playing field and cannot be called fair trade.

If the breakthrough patents in life forms are any indication, this field will increasingly become the domain of large corporate concerns, not universities who develop the technologies and not smaller companies which would constitute a level playing field. The first patent granted for a living organism was for the oil eating bacteria isolated by Ananda Chakravorty at Chicago. The patent is held by General Electric. The famous oncomouse developed at Harvard is "owned" by Du Pont which holds the patent on these controversial animals (18, 19). Now we are told that Sandoz posesses a mouse model for Alzheimers disease; whether they will patent the animal or hold it in seclusion for themselves, remains to be seen.

Consequences for Developing Countries

The consequences of patenting genetic resources would be highly detrimental for developing countries. These are primarily agricultural economies and therefore heavily dependent on genetic resources for upgrading and maintaining agricultural stocks. Privatisation of genes, leading to license fees for their use would not only result in the absurd spectacle of developing countries being required to pay for the use of their own property but also increase the already strained research budgets in such countries. Scientific research would suffer, leading to diminishing technological capacity, exacerbating the technology gap and increasing dependence.

Althought patents and PBRs have a "research exemption" clause which allows protected material to be used for basic research, the results of such research can not be applied commercially. Developing a technology means commercial application of research results. This would not be possible without the payment of heavy license fees for using patented genetic material. This would for example mean that Indian scientists would be restrained in their ability to use plant germ plasm to breed improved varieties for their farmers.

It would also mean that India would be hindered by patents to develop an indigenous biotechnology industry using its germ plasm. If the compounds in the Indian Neem are patented by a foreign company like WO Grace, Indian industry would have to pay royalty to be able to establish its own Neem based biopesticide units. If, for example the characteristic of frost resistance is patented, as is being attempted by some companies (20), breeders in developing countries would be denied access to this greatly advantageous property, even if it was induced through a route, or using genes different from the original. That is what happens with product patents.

The promising toxin gene in Bacillus thuringensis, being used to build in insect resistance into plant varieties, has been patented recently by Plant Genetic System of Belgium (21). It could mean that other breeders would not have free access to the Bt gene for any other purpose. Such monopolies cannot possibly bode well for the progress of science and for all round development, in developing countries, or industrialised countries either.

Genetic Erosion

Ecologists and biologists are concerned about another dangerous offshoot of gene patents. That is genetic erosion. The pro patent lobby argues that patenting is not the only process to induce genetic erosion, that intensive cultivation, particularly the Green Revolution has already contributed to that. That seems a very silly argument. Just because other systems have already caused damage, does not mean the way is opened to compound the problem still further.

"The ultimate danger of increased reliance on corporate r – DNA crop regimes is that eventually there will be few alternatives to genetically engineered seed. . . Farmers who want to buy open pollinated seed will find it increasingly hard to do so. Consequently, the current trend of farmers switching to ecological methods of farming, like permaculture, organic and biodynamic farming, could be seriously retarted." (Richard Hindmarsh, Griffiths Univ., Australia)

Patents and other forms of privatisation will lead to an even greater narrowing of the genetic base. Plant varieties carrying a few economically important traits like high yield or resistance to a particular disease, have been and will continue to be marketed aggresively by the patent or PBR holders. This leads to only few varieties being grown for a crop, displacing older varieties and resulting in the loss of their genes. This will have very unfortunate consequences for the viabilty of our agricultural stocks and their ability to survive biotic and abiotic stresses.

The chemical seedsmen, on the other hand, are intensifying the chemical approach so as to combine their traditional sales of pesticides with their new interest in seeds. Their special corporate approach is to engineer seeds so that they carry inbuilt genetic tolerance only to the pesticides that they produce, and to no other. Therefore, such corporations acquire twin monopolies, that on seed through patents and on their herbicides, which will be the only brand that their seeds will tolerate. Wherever they are able to push their seed sales, their pesticides will also have to sold, otherwise their seeds would fall prey to disease. The combined effects of this approach will be ghastly.

"Worldwide, over 79 corporate/state research programs are developing more than 23 herbicide – tolerant crop lines, including cotton, maize, corn, potato, rice, sorghum, soyabean, wheat, tomato, alfalfa and surgarcane. The most likely result will be a greater entrenchment of the chemical approach, which in turn will further increase soil and water pollution, pest resistance and chemical residues in food. In the process, natural ecological processes will be further distorted and diminishing biodiversity will worsen." (The World Herbicide Front)

Furthermore, as the race for patents intensifies, and the argument of faster and more efficient returns on investment gains ground, it will increasingly be patentability of products that will determine the thrust of research, not human need. That will be an unfortunate developed indeed.

Biofertilisers and Biopesticides

In modern agricultural practice, there is despite vested corporate interest, an attempt to move away from chemical fertilisers and pesticides. Intensive, yield oriented agriculture made possible by heavy inputs of chemical nutrients and high doses of pesticides has certainly helped to overcome hunger in many parts of the world, but it has also extracted a heavy price. The poisoning of our land and water has brought home to us in an acutely uncomfortable way that agrochemical use will have to be curtailed if the human race is to continue to feed itself.

With this in mind, there is a greater impetus to develop benign substitutes for agrochemicals. Although they do not dominate the market yet, biofertilisers and biopesticides will have to play a much larger role in agriculture. They are already being used with success in some parts of the world. Cyanobacteria and Azolla have shown promise in rice cultivation and other water plants, algae, and micro organisms are being put through field trials (22, 23). Similarly, natural enemies of pathogens are being rediscovered and the Bt model of building resistance into the plant itself is being tried with a variety of other insect toxins (24). By virtue of being based on living organisms, the entire spectrum of biofertilisers and biopesticides would come under the purview of patents and with that become subject to monopoly control by patent holders.

Scientific research is unfortunately as much a matter of money as of talent. Once the starting line is made the same for every one, the advantage to patent will automatically lie with the industrialised countries, and very particularly with the corporate giants in these countries who have large amounts of resources to invest in scientific research and development.

And lastly, with biotechnology, it is for the first time since the industrial revolution that developing countries, or atleast some of them, can be participants in a technological revolution. This is because biotechnology is in a very fundamental way unlike other technologies which have in their time, been money spinners.

Success stories like electronics and informatics are hugely capital intensive, which biotechnology is not. It is rather a skill intensive technology where a major portion of the investment is in the form of trained manpower. This is also the reason why many developing countries, which are acutely cash strapped but have reasonably well developed human resources in the form of skilled manpower, could become co-players in biotechnology.

This technology offers developing countries, provided the field is level, the opportunity to engage in self-reliant economic growth. Some of the economic imbalances could be corrected if developing countries could participate in biotechnological production methods. Building local technology capacity based on genetic resources, a material that people, specially farmers in developing countries are familiar with at a fundamental level, would lay independent economic foundations in countries that have so far been unable to climb out of the morass of poverty and economic dependence.

Countries like India, China and Thailand in Asia; Brazil and Venezuela in Latin America, and Egypt in Africa, have invested in biotechnology in an attempt to achieve economic self reliance. These efforts would suffer a setback if IPR systems completely disparate with their requirements were enforced in these countries.

Table 2. Biotechnology Thrust Areas – India

1. Agriculture – seed, biofertilisers, biopesticides and microbial inoculation
2. Plantation crops, spices, forestry and agriculture
3. Vermiculture
4. Animal husbandry-vaccines and diagnostic kits
5. Aquaculture
6. Sericulture and leather biotechnology
7. Products for human health
8. Industrial biotechnology – alcohol and alcohol based feedstock chemicals and antibiotics
9. Basic research

To judge the possible costs of IPRs on agriculture, the mainstay of developing countries, the World Bank had commissioned a study. The main points of the report released in 1990 are shown in Table 3.

Table 3. Effects of IPR in Agriculture. (World Bank)

1. Private sector breeders shift focus of research to patentability and commercial value; not human need.
2. Inhibition of free flow of information between and among public sector.
3. Lower federal research funds, affecting orphan research areas.
4. Genetic erosion.

In developing countries, plant breeding and seed production are in the public domain, financed with public money and therefore not protected by IPRs. There is a good reason for this. In a country like India or Malaysia where farming is the major occupation, the agricultural community is composed largely of small and marginal farmers practicing subsistence agriculture. Their requirements for reasonably priced seed of minor, of neglected crops are fulfilled by public breeding institutions. The private sector with its dependence on the market for survival is thought to be risky as a source of seed for poor farmers dependent on orphan corps.

In addition, in India, the field of seed production should logically be used to generate employment in a decentralised way, for the rural population. Rural youth who lack other technical skills could be profitably employed as the agents for generating certified seed from foundation seed and play a useful role in producing seed from asexually propagated plants like potato. It is with these intentions that in India, seed production for foodgrains has not been given over to the commercial sector which is however active in horticulture and floriculture. The cost of introducing IRPs in agriculture would be decidedly detrimental to the small and marginal farmers.

The collapse of a very European experiment in the socialist model of government and economy has reaffirmed the belief in Europe and the US that the market economy with its overwhelming reliance on the private sector and its rejection of the state is the only sensible way to structure the economy in all countries of the world. This view is not only terribly simple, it is also incorrect.

Free enterprise is only justified when all members of a society have atleast theoreticaly equal chances to participate in the economy. This is not the case in most developing countries where a very large section of the population is poor and uneducated. In their present state of development, they would have absolutely no chances of participating in a free market economy and therefore their areas of economic interest, like agriculture and seed production, must be protected to some extent by state intervention. Promoting a pure market economy is as much against the interests of the economically vulnerable sections of the Third World as imposing western models of IPR.

An Equitable IPR

I am totally opposed to the concept of "harmonising" Intellectual Property regimes across the globe. I am of the view that each country must have the right to develop an IPR system that suits its requirements and is in keeping with its stage of technological development. I have listed in Table 4, what I consider to be the five points of an equitable IPR regime for India and other developing countries.

Table 4. Equitable IPR Regime

1. Exemption for farmers
2. Exemption for breeders
3. Acceptance of farmer's rights
4. Payment for genes as for any other natural resources
5. Genes from international gene banks to be excluded

1. The farmer's privilege to produce seed for himself, as well as his right to sell and modify seed after paying the initial licences on protected seed must be retained.

2. The breeders must continue to have the right to use the protected genetic material to breed other varieties. This should be distinct from the so called research exemption which allows research on protected genes but not their commercial exploitation.

3. The Farmers' Right, as distinct from the aforementioned farmers' privilege will have to accomodated in the IPR system. This right is acknowledged by the FAO which has passed a Farmers Rights bill in 1987. The Farmers Right states that the informal innovation of Third World farmers over generations, in creating and maintaining the world's germ plasm, will have to be placed on par with the more formal innovations of modern plant breeders. If one is to be

compensated by breeders rights, the other will also have to be rewarded for his labour and creative genius.

4. Genes which have always been exploited commercially in agriculture, have now become a veritable commodity in trade, with the growth of biotechnology. If they are to be subjected to any form of privatisation, they must be paid for like any resource that is used as a raw material to create products. Surely no right thinking person will hold a brief that genetic resources should continue to be common property which do not have to be paid for, and at the same time press the case that they should be patentable.

5. A final condition should be an agreement that genes stored in international gene banks like those of the CGIAR system (Consultative Group on International Agricultural Research) should be kept out of the purview of farmers, overwhelmingly from the Third World. They were taken from fields to be kept in trust to ensure global food security. The mandate of the CGIAR system is not to provide raw material to multinational companies, it is to create conditions to feed the human race.

In Table 5, we see listed some of the anticipated costs of patents for developing countries. Apart from the formal heads like higher administration costs, and increase in the cost of research, the increase in royalty payments would lead to a big drain on the economies of these countries (25). Since IPRs are sought to being introduced into developing countries through the GATT trade round, in GATT terms, royalty payments would amount to 5 % of global trade! Since developing countries are technology importers, the bulk of this would be paid by them, to the industrialised countries. The foreign exchange outflow from the Third World would be so immense, as to cripple effectively, any attempts at sustained or self reliant growth in these regions.

Table 5. Cost of Patents for India

1. Administration and enforcement costs.
2. Increased Royalty Payments –
   a. Developing countries net importers of technology,
   b. Royalty payments in GATT could amount to 5 % of global trade,
   c. Foreign exchange outflow from third world could double or treble.
3. Higher costs of R & D, reduced research opportunities
4. Restricted local technology capacity.
5. Monopoly on market and high prices.
6. Shearing of compulsory licensing and treating importation as working free technology dissemination.

The consequences of such developments will be grave indeed. In the developing countries, uncontrolled poverty and indications of foreign domination of the economy will lead to large scale unrest. The resulting turmoil and political instability cannot bode well for democratic institutions and will automatically create the conditions for extreme political configurations. This ist not a desirable development anywhere.

Moreover, it cannot serve corporate interests to so over reach themselves that the conditions in the countries where they want shares of the market, become unstable and hostile. The selling of patented products is only possible in a stable, functioning economy. I would like to call to your attention here the developments in the late 70s when the Coca Cola company was asked to wind up its operations in India since the Indians felt that the way the company was functioning was not in the country's interests. There is a lesson to be learnt here.

As a scientist, a member of a developing country and a global citizen committed to the belief in a just and equitable world order, I would like to use this platform to issue a warning in the interests of all of us. If this heavily skewed, exploitative, unjust model of developement discriminating between the rich and poor is not righted, the rage and frustration of the hopeless, hungry masses will pull this edifice down.

For those sitting secure in their belief that nothing can touch their affluence since there are thousands of miles and oceans between the haves and have nots, let me point out that television has brought into every hungry home the wasteful consumption of the west. The desperately hungry have nothing to lose. And certainly not when awareness dawns that it is the rank injustice, the plunder of their own property that has created the chasm between their emaciation and the plump prosperity in far away western countries. The knowledge that 20 % of this world consumes 80 % of its resources and spews 80 % of pollution into its land, water and air, is being less and less tolerantly received.

If there is no serious, visible effort to rectify these wrongs, then we will see migrations across continents in the search for food and jobs. You who are sitting here have seen the boat people from Vietnam, you have seen the waves of asylum seekers knocking at your doors. That is only a foretaste of what could happen if the present trend of still greater disempowerment of the South continues. Surely people like you, influential members of the scientific, legal and economic communities, members of governments and concerned citizens, would see the need to send the right signals. This is the challenge and the responsibility of the intellectual class of the world, the people who should and do guide policy, to use their influence to see that development in the world is just and equitable.

References:

1. Vavilov NI (1926) Studies on the origin of cultivated plants. Bulletin of Applied Botany, Genetics and Plant Breeding. 16:139–246
2. Moony PR (1983) The Common Bowl. Development Dialogue 1–2:7–23
3. Prescott AR, Prescott AC (1982) Wildplants and Crop improvement, World Conservation Strategy Occasional Paper No. 1, World Wildlife Fund (undated) and What is Wildlife Worth? Earthscan, London, 1982

4. Dolan DD, Sherring W (1982) Progress in the evaluation and use of plant germplasm in the Northeast (1974–1979) Special Report No 43. New York State Agricultural Station, Geneva, NY
5. Murray JR, Alexander H (1981) Biological Diversity and Genetic Engineering, presented to the State Department Conference on Biological Diversity, November 1981
6. Kendrick JP Jr (1977) Preserving our Genetic Resources, Californian Agriculture. September, pp 2
7. International Undertaking on Plant Genetic Resources (1983) Resolution 8/83, 22 Nov C/83/Rep/8 Rome. FAO
8. Brockway LH (1979) Science and Colonial Expansion. The Role of the British Royal Botanic Gardens. New York Academic Press
9. Prescott AR, Presscott AC (1982) Wildplants and Crop Improvement, World Conservation Strategy Occasional Paper No 1, World Wildlide Fund (undated) and What is Wildlife Worth; Earthscan, London
10. Bienz TB (1993) Biopharmaceuticals go to market: Patterns of Worldwide development. Bio/Tech 11:168–172
11. Gupta R (1993) Emerging medicinal plants in world Pharmaceutical Industry. Indian Horticulture, April–June:43–46
12. Raundal A (1991) The value of Biodiversity Ambio 20:64–68
13. Seedling, November 1992, pp 11–14
14. Stone R (1992) A biopesticidal tree begins to bloom. Science 255:1070–1071
15. World Grain Situation and Outlook: Circular series. FG 6-93, June 1993. United States Agriculture Department
16. International Convention for the protection of new varieties of plants, December 2, 1961; as revised at Geneva on October 23, 1978
17. International Convention for the protection of new varieties of plants, of December 2, 1961; as revised at Geneva on March 19, 1991
18. US Supreme Court (1982) Diamond commission of Patents and Trademarks Vs Chakrabarthy. United States Reports 447:303–322
19. Nature (1988) 332:668
20. Lindow SE (1990) Use of Genetically altered Bacteria to achieve Plant Frost Control. pp 185–210 in Biotechnology of Plant – Microbe Interactions. (Eds Nakas JP and Hagerdon). Mc Graw – Hill, New York, USA: XI + 425 pp
21. European Patent Office (1993) European Patent No 0451878
22. Hindmarsh R (1991) The flawed "sustainable" promise of Genetic Engineering. The Ecologist 21:196–205
23. Ibid. pp 198
24. Dixon B (1987) The Gene Revolution. The UNESCO Courier, March 13–16. Postgate J (1990) Fixing the nitrogen Fixers. New Scientist, 3 Feb, 57–61
25. Whitton BA, Roger PA (1989) Use of Bluegreen Algae and Azolla in Rice culture in Microbial inoculation of Grop plants, R Campbell and Macdonald (eds). Special publications of the Society for General Microbiology, 25, IRL Presse, Oxford, England, 89–90
26. Boulter D (1981) Genetic Engineering of Plants for Insect Resistance, Outlook on Agriculture 18, 1, 1989 pp 2–6. Boulter D, Gatehouse JA, Gatehouse AMR, Hilder VA (1990) Genetic Engineering of plants for Insect Resistance. Endeavour, New Series, 14 pp 195 illustrated
27. Siebeck WE (ed), Evenson RE, Lesser W, Braga Carlos AP (1990) World Bank Discussion Papers 112
28. Ibid

Patenting of Human Genes and Living Organisms: The Socio-Political Situation in the United States of America

B. I. Rowland

Molecular biology has made tremendous advances in its ability to understand and manipulate biological processes, both *in vitro* and *in vivo*. The capability to change life forms, cells, plants and animals, has aroused many concerns: religious, environmental, economic and sociological. As part of these concerns, attention has been directed to the patenting of inventions associated with these new biological advances. As will be discussed, under U.S. law as it is understood at present, most, if not all, molecular biological inventions will fullfill the statutory requirement of being patentable subject matter. Therefore, under the laws of the United States, there is no impediment to filing for inventions in the field of molecular biology, regardless of the nature of the particular invention and obtaining a patent, assuming that the other requirements for patentability are met. The types of inventions in molecular biology extend over a broad ambit, involving compositions such as nucleic acids and proteins, genetically modified plants and animals, methods of producing these products, and methods of treating animals, including humans. Each of these categories of inventions has raised varying degrees of concerns, the least of which being the compositions.

Patenting of Human Genes

Patenting of human genes is well established today. (There is a large patent literature where numerous genes have been patented and in a number of cases commercialized.) The number of patents claiming genes is extensive. In a number of these patents, the patented genes have been commercialized. Human growth hormone, insulin, erythropoietin and tissue plasminogen activator are only a few of the proteins which have been patented and which are commercially available for therapy today. There is, therefore, strong economic support for patenting human genes and pure proteins, where the protein can find use in therapy, in identifying drugs which may be used in therapy, or in understanding physiological processes, which may then allow for therapies.

Furthermore, the manner in which the products are made, fermentation, is an ancient practice. Fermentation is carried out within factories, which are highly controlled, and are not likely to affect the environment in which they are present. Therefore, they do not intrude into the environment, nor do they create significant concerns about their effects on the environment or the people involved with the manufacture.

The patenting of these compositions is becoming more difficult, as the means for isolating, identifying and purifying these compositions have greatly advanced and been simplified . Therefore, as to any particular gene or protein, there must be less and less known about the composition, in order to support patentability. In light of the enormous investments made to identify genes and proteins which may have therapeutic use, there will continue to be a strong interest and industrial support for their continued patentability. It will only be where they are used in certain applications, such as gene therapy, that one might find concerns with, and hostility to, their application.

Patenting Amimals

So far as patenting animals in the United States, while neither the United States Supreme Court nor the Court of Appeals of the Federal Circuit which reviews all patent matters, have not specifically ruled on this issue, the general belief is that animals will be found to be patentable subject matter. With the decision of *Diamond v Chakrabarty*, 477 U.S. § 303 (1980), the Court held that live human-made microorganisms were patentable subject matter. The Court used sweeping language in providing that anything under the sun made by human hand could be patentable subject matter. This decision opened the way to numerous applications directed to living organisms. On the way to finding animals patentable, the Patent and Trademark Office Board of Appeals found plants patentable in Ex parte *Hibberd*, 227 U.S.P.Q. 445 (DTOBPAI 1985). In this decision they found that there was no impediment to the patenting of plants due to the other statutory basis for obtaining plant patents. The Board was not particularly concerned with the fact that a plant is a multicellular organism which is much more sophisticated than a microorganism, but rather looked to whether the Plant Patent Act ("PPA") and the Plant Variety Protection Act ("PVPA") exclusively controlled the manner in which plants could be patented. Having accepted that multicellular organisms could be patented, the Board did not find it too difficult to also find that animals are patentable when confronted with this issue directed to polyploidy oysters in Ex parte Allen, 2 U.S.P.Q. 2d 1425 (1987). Using the sweeping language of *Chakrabarty* for support, the Board held to that so long as the hand of man had created the living animal, there was no reason to distinguish living animals from other living organisms for purposes of patentability.

Despite the approval for patenting animals, very few animal patents have been issued by the United States Patent Office. The patent that has had the greatest notoriety is referred to as the "oncomouse". This patent is directed to a mouse which has greater susceptibility to cancer and, therefore, finds use as a model in cancer studies. Since the issuance of the oncomouse patent, three other patents to animals have issued. The Patent Office takes particular concern in issuing patents to animals, providing for a much more intensive review of the application, prior to issuance.

The question has been raised whether it is moral to patent animals. The law does provide that immoral inventions may not be patented (Lowell v. Lewis 15 F. Cas. 1018 (No. 8568) C.C.D. Mass. 1817). It is, therefore, a legitimate question as to what was the intent of the statute, but the general attitude has been that patenting is not a moral issue. Rather, the concern is whether the act is moral, namely genetically modifying the animal, rather than whether the patenting of the act or the result of the act is moral. Morality only becomes involved in that patents enhance the incentive to modify the animal. However, unless society finds that the act is immoral and should be outlawed, there would appear to be little basis for denying patents for genetically modified animals.

There are also the religious considerations that man is modifying what God created. One might argue that genetic modification of animals is just a difference in degree from breeding animals, since breeding also results in genetic modification and directs, by human intervention, changes in what God created. However, breeding is a slower and less precise process with much fewer opportunities for modification than genetic modification using molecular biological techniques. Therefore, modifying God's plan as to how humanity is to develop through natural breeding raises fewer concerns about intervention with the "divine plan".

Litigation

The Court of Appeals of the Federal Circuit had an opportunity to consider whether animals were statutory patentable subject matter in Animal Legal Defense Fund vs. Quigg, 932 F.2d 920, 18 U.S.P.Q. 1677 (Fed. Cir. 1991). The Animal Legal Defense Fund sued the Commissioner of Patents to have the Commissioner enjoined from issuing patents to animals. the complaint was predicated on the Administrative Procedure Act ("APA"), that the statute required that there be public notice and comment prior to the issuance by the Commissioner of the notice that the Patent Office would consider animals as statutory subject matter. A number of individuals and groups joined with the Animal Legal Defense Fund to challenge the Commissioner's notice. The challenge raised a number of different issues.

The first issue that was raised concerned the requirements of rule making. The solicitor for the Patent Office argued that announcing that animals were statutory subject matter was not rule making as provided for in the APA. Rather, the court found that it was merely interpretive of the law and under the exception in the statute which excludes interpretive rules, the Commissioner was not required to publish the notice for comment, prior to its promulgation. The Court of Appeals affirmed the lower court in its dismissal of the action. However, the plaintiffs had raised other issues, which had not been considered by the lower court. The Court of Appeals *sua sponte* considered these issues as well.

A second count alleged that the Commissioner had exceeded his authority in providing for animal patents, since Congress was required to act. In order for the plaintiffs to be able to challenge the Commissioner's act, it was required that they have standing. By "standing" it is intended that they would suffer an injury if the Commissioner's Act was not set aside. Among the alleged theories was the theory that if patents to animals were granted, farmers would be required to increase their cost of operations, due to the paying of royalties for the patented animals. Furthermore, increased productivity from the animals would have a negative impact on small farmers who could not compete.

While for the purposes of the appeal, the injury was admitted, the solicitor argued that there had been no showing of causation, namely, that the Commissioner's allowing patents on animals would result in these injuries. The court found that the plaintiffs did not support the requirement of showing causation, since there would have to be intervening third parties, whose actions could not be readily predicted, and that the granting of the requested action would probably not redress the grievances alleged. This count was also dismissed.

Finally, the court found that the APA required the party complainant to be within the zone of interest of the statute. The court found that the patent laws were not directed to the interest alleged by the complainants and therefore, the complainants failed to have standing. Thus, there was no ruling on whether animal patents are valid, but rather the court avoided the issue by dismissing the suit on procedural grounds, without any comment on the substantive issues raised.

While it is clear that the Patent Office deems that any animal, other than humans, may be patented, this has not been required by any decisions of the Board of Patent Appeals. One can make a distinction between polyploidy oysters and higher animals. Issues associated with higher animals, particularly laboratory animals, are not associated with cold blooded animals, particularly shellfish. One could, therefore, make a distinction between animals with sophisticated nervous systems and animals which lack sophisticated nervous systems. Nevertheless, no one has raised this issue as yet and we still await a definitive decision from the reviewing courts. By the time the issue is raised in court, there will then have been a very substantial number of patents issued claiming genetically modified animals, as well as a very large amount of investment in pursuit of inventions associated with the patenting of animals. The court, therefore, will be faced with an existing system which will have accommodated itself to the existence of animal patents. Under these circumstances, unless some horror occurs to shock the Court concerning the role of animal patents, it is extremely unlikely that they would find that animals are not statutory patentable subject matter.

Legislation

Besides attempting to stop animal patents by litigation, those against genetic modification of animals have looked to the legislature. A number of Congressmen have attempted to restrict the scope of patenting of living things, both plants and animals. The issues are quite different between plants and animals. For plants, the issues are primarily economic; for animals, the issues are primarily moral.

At hearings held by the House Subcommittee on Courts, Civil Liberties and the Administration of Justice, Representative Charles Rose, Democrat of North Carolina stated, "this patent policy makes no distinction between living things and inanimate objects." The animal kingdom is redefined in technological terms to "suit the commercial needs of the marketplace." He indicated that he feared humans would lose their common heritage as animals and become the "sole possession of major corporations." He declared, "we are playing God, and must not allow the marketplace to establish the future of the animal kingdom."

In contrast to Congressman Rose, Dr. Ann Sorensen of the Farm Bureau testified that "new biotechnologies may eventually enable us to reduce farm costs, improve the environment, and expand the utilization of farm products. . ."; Dr. Sorensen further commented "[t]he transfer of genes will confer increased growth performance and disease resistance, as well as various beneficial reproductive traits, which collectively will lower input costs and produce a more healthful product for the American consumer."

Another view was put forth by Mr. Cy Carpenter, President of the National Farmer's Union. He argued that Congress should assume the responsibility for the determination of patentability of animals. He compared the situation with the seed industry, where patenting was allowed. Since the allowance of seed patenting, the acquisition of independent seed companies has accelerated. He indicated that presently five major corporations control 120 formerly independent seed companies.

As might be anticipated, William H. Duffey, a Director of the Intellectual Property Organization speaking on behalf of both IPO and the Industrial Biotechnology Association ("IBA") supported the patenting of animals, indicating that the U.S. had become a model for biotechnology patents, that the issuance of patents was morally neutral and that it did not displace what already existed in breeding. In a subsequent hearing, Richard D. Godown, President of the IBA argued for the potential substantial improvement in animals in making them disease resistant. Furthermore, he argued that if the United States didn't provide for the genetic engineering of animals, U.S. farmers would be forced to compete with animals in other countries which had been genetically modified. Finally, Mr. Godown stressed that transgenic animals would come within the extensive regulations by the United States already governing animals today.

There have been two major proponents of legislating restrictions on living organisms in the Congress. Senator Hatfield of Oregon has been the most aggressive in his efforts to restrict the patenting of animals, repeatedly introducing statutes for a moratorium on animal patents. While one of these bills successfully passed both the Senate and House, it was killed in committee hearing. That ist the farthest any bill went in the legislative process and that success has not been repeated. The other approach has been Congressman Kastenemaier's, whose bill was directed to the patenting of plants, but contained notions applicable to animals as well. The Kastenemaier bill as directed to animals would provide that genetically altered animals are patentable subject matter; human beings are not patentable subject matter; the Commissioner is authorized to issue any regulations necessary to regulate the deposit of biological materials; and that, "it shall not be an act of infringement for a person whose occupation is farming to reproduce a patented transgenic farm animal through breeding, use such animal in the farming operation, or sell such animal or the offspring of such animal." However, "notwithstanding the provisions of paragraph (1), it shall be an act of infringement for a person to sell the germ cells, semen or embryos of a patented transgenic farm animal." This statute has aspects of the PVPA. However, opponents of the Act pointed out that there are substantial differences between the situation concerning plants and the PVPA and genetically modified animals. As of this date, this statute has not been passed and appears not likely to pass. Therefore, at the present time it is unlikely that there will be any restriction on the patenting of animals, and that any restriction will be associated with regulation of animal care.

Laboratory Animals

The genetic modification of laboratory animals can lead to chimeric or transgenic animals. For the purpose of this discussion, chimeric intends the introduction of a xenograft into an animal, where the xenograft is able to survive and function. The term "chimeric" is also used to describe animals where a gene has been introduced into an embryonic cell and the resulting animal is a mixture of the wild-type cell and the modified cell. This type of chimera will be included under transgenics.

The chimeric animals have for the most part been mice into which human tissue has been introduced, particularly normal tissue. A large body of work has developed around the *scid/scid* mouse and, as further genetic modifications provide for a variety of other immunologically compromised mice, these animals will also find application. The advantages of having human tissue are evident, where one is concerned with the effect of agents on human tissue. This is graphically exemplified by HIV, which is tropic for human cells. Therefore, as to those diseases where the pathogen is tropic for a human cell, where one is concerned with the effect on the physiology of a human cell or organ, or when one wishes to study human tissue in isolation, the immunocompromised mouse affords a convenient environment. Proponents of the use of these animals point to the unique advantages that these chimeric animals provide for studying human disesases.

For the most part, the human tissue which is used is fetal tissue, which appears to be able to grow much better in the chimeric animal than adult tissue. The chimeric mouse confronts animals rights and anti-abortion activists. Neither of these groups have wide-spread support today, but both are vocal and their support may grow. However, until either or both of these groups obtain substantially increased public support, chimeric animals will find wide spread application in the laboratory.

The transgenic animal is a result of genetic modification by modifying an embryonic cell, either oocyte, usually by injection of DNA, or an embryonic stem cell, using a variety of techniques for introducing the DNA. Since the oocyte must be fertilized, the oocyte is a heterozygote. In the case of embryonic stem cells, only mouse embryonic stem cells have been successfully used. The procedure provides, where there are two copies of the gene, only one copy is modified. However, there are now techniques for modifying both copies. In any event, the modified embryonic cell is introduced into a blastocyst with the result that one obtains animals having wild-type cells and genetically modified cells, where one or both of the genes have been modified. In both cases, one can then breed the resulting animals to obtain homozygotes, where the animal comprises only those cells that include the genetic modification.

Of particular interest for these types of animals is the process of homologous recombination, where rather than having random integration and providing for transcription or expression of an additional gene, one can directly modify a specific gene in the host. This has been exemplified by "knock out", where one inactivates a particular gene or may provide for expression of a different gene, which may have equivalent capability. For example, there have been extensive efforts to modify the immunoglobulin locus, so that a human immunoglobulin locus would supplant the mouse immunoglobulin locus. In this way, antibodies produced in response to immunization or other immune stimulus would have the human protein sequence, rather than a mouse protein sequence. Mice have also been produced in which the β_2-microglobulin gene has been knocked out, so as to substantially compromise Class I major histocompatibility complex antigen dependent processes. These animals find use in identifying immune responses to a variety of pathogens, as well as the response to cancer. Genpharm, a small biotechnology company, is developing these animals and has already offered some of these animals for sale.

In the case of larger farm animals, genetic modifications have been introduced which allow for production of human proteins in milk. In this way, cows can be produced where milk may be fortified with specific proteins which may have important physiological activity in the gut. Alternatively, the suggestion is that they be used as commercial factories for producing therapeutic proteins which may be glycosylated or otherwise processed analogously to human proteins. In addition, the suggestion is that by genetic modification, beef and pork may be produced which is leaner, animals produced which are more disease resistant, protein produced which provide for an improved human diet, as well as other desirable characteristics. Commercialization of these improved patented animals should not be too far away.

The patenting of domestic animals in clearly a concern to small farmers. While it is true that one cannot foresse the future and how companies who develop genetically modified animals will act, one scenario is that they will buy up farms to stock the farms with these animals, rather than license them generally to existing farmers. The farmer's argument that the patented animals would be more costly is not convincing, because without economic advantage, nobody will use them. The real issue is whether the economic advantage will be so great that large agribusiness will dominate the use of these modified animals to the detriment of the smaller farmer. Such activity is made feasible only by the patent system, since trade secrets are not likely to suffice. Therefore, there is a social interest in the patenting of domestic animals and how those patented animals are used. This does not mean that the patenting of domestic animals should be precluded. Rather, once it is seen how the industry uses their patent rights, legislation can be devised to regulate such use in accordance with societal interest. Such legislation can be designed to continue to provide incentive to modify domestic animals, so that society as a whole can benefit, while at the same time regulating how the domestic animals are commercialized, so as to accomodate other societal interests.

Executive Support

Biotechnology enjoys enormous executive support. Vice-President Albert Gore has always had an interest in the biotechnology industry and has repeatedly indicated that he believes this industry can provide jobs and income for the United States. Where heavy industry is reducing the total number of jobs, government is increasingly looking to the new technologies, of which biotechnology is one, to take up some of the slack. While it is unlikely that biotechnology can provide the number of openings which are necessary for low unemployment, nevertheless it can be a significant contributor. By providing for high paying jobs, biotechnology can create a group of consumers, which will benefit a large number of other workers.

In addition, one of the greatest concerns in society today is health care. The United States is particularly concerned about the rate of increase in the cost of health care and its impact on the economic state of the United States, unless the run-away costs can be turned around. One facet of concern is the search to provide cheaper forms of treatment, prophylactic and therapeutic. The biotechnology industry offers enormous opportunities in this area, where the biotechnology industry can realize significant profits, by substituting new therapeutic modalities which are far less costly, for present therapeutic modalities, which may involve expensive hospitalization, repeated visits to doctors' offices, or other high cost events.

Finally, the United States has been a prime mover in molecular biology. There is an enormous amount of national pride in what has been achieved in this field. To that extent, society is willing to overlook its concerns about genetic engineering of plants and animals, in the absence of a "clear and present danger."

In conclusion, patenting of biotechnology is part of the entire biotechnology scene and enjoys strong support from a wide variety of communities of interest. While there will be challenges, they are unlikely to be significant or to succeed.

Discussion: Session 5

Hofschneider: Dr. Sahai, thank you very much in particular for the last words you said for the morning I think we have to discuss it seriously. On the other hand, I had hoped after the two presentations that we would be able to compare the socio-political situation in the U.S. and in the Third World, in particular India, regarding the human genome, the animals and the plants, but now we have the situation that you only discussed plants and Dr. Rowland only discussed animals and humans. Now, do you want to add a few words to the plant situation after having heard the second contribution, before we go into the discussion?

Rowland: I think my title was human genes and animals, so I didn't address any issues. My only experience in plants was working with Calgene which was one of the early biotech companies. Their direction was primarily toward commercial high price products such as tomatoes, they have what they call the flavor saver which, unfortunately, while it travels very well doesn't taste very good. They were not interested in other crops, they're not a seed company. On the other hand Pioneer which back in the 80ies was not interested in biotechnology has become quite heavily involved and they are a large seed company. The problem I have is if they exploit the various gene banks that are available and they make that investment in providing for products. As you very rightfully said, a product that may work very well in the United States may not be at all suitable to India. They will have to develop plants specifically for the Indian climate and Indian needs which requires investment and the question then is, what is to encourage that investment? And I also felt that in your remarks you point out that Queensland quite properly made it a national and natural heritage which is their own property and that many of the issues you raised could be done nationally, but India could also say that any of their flora and fauna are their property and if one wishes to use it one would have to pay for it. I think that would be very legitimate, but it's not the West's problem, it's India's laws which have to be passed. To the extent that they have not been passed, I guess at least the governments I work with are going to take advantage of it. So, as far as plant patenting is concerned I think it's a very important area, I think they're going to first modify plants with very large profits and that will certainly mean that they will not be interested in trying to develop crops which cannot be easily adapted from a Western crop to an Eastern crop. I don't see a change in that.

Sahai: The reason I concentrated on plants was because that's of primary importance to us. Of course, all that is happening with human genes would have consequences for us in terms of closing our options or making the options more difficult. Therefore, it is not a step that would be very welcome but it is something that, if patents for all of the other reasons are to be extended on human genes, I have objections from the scientific point of view and I did want to go into what Dr. Zimmerli had said earlier. I possibly want to take that up in the round table dicussion. I think I would be grateful if we could concentrate the discussion right now on what we have said rather than open new avenues.

Beyreuther: I was very impressed by your comment on the patenting of animals, but we always have to think of the worst scenario and the real animals like the one for Alzheimer's disease or other major disorders have not been produced and these may be very unique animals. The costs for patenting an animal have to come out of the product you get with this animal and the two things I think may come up in the near future are those mice that you mentioned – the knock-out mice – it may be used to produce a novel type of antibodies. If you knock out a gene, then the animal does not become tolerant towards this gene and then you may get very potent antibodies in those animals that can be used for medical purposes or any other detection purpose. Each housewife may use it to test whether the food in the fridge is still edible or not and then the animals may come back into play and maybe become very important products of modern industry.

Rowland: If we're talking about laboratory animals there's a lot of work going on with multiple sclerosis, there's a study where they put neural tissue into the eye of a mouse and it turns out that you do get a blood/brain barrier; and they are starting to work on HIV, so these animals are very important, but I just don't think patents will be important to them, I think they'll come out of the universities, they'll be directed to the various non-profit institutions that provide animals today. As far as domestic animals are concerned I think they'll be improved, particularly disease resistance is very important. So, I think the investment is very wise and we are investing in all these areas and they're also developing models, so I agree with you that this is a very important area, but I don't think patents are going to run the laboratory animal side.

Beyreuther: I agree with you fully but you still say that if they are worth the money we are going to patent them and I think that's something we should not do from all we know right now. Why can't we just make this statement? You, as a person who deals with these kind of issues.

Rowland: Let's take Systemix which has developed the SCID mouse, I would say they put about 8 to 10 million dollars into that animal and I could unterstand their reluctance to make it publicly available without getting some quid which could be technology, it needn't be money. As far as the other animals are concerned, I really don't unterstand why MIT licensed it to Genfarm and it turns out the other animal is licensed to Cell Genesy but my suspicion is that the animal will end up in these places. There's a different situation where they want to make human antibodies where they'll transfer the gene for that whole locus and that is a very major scientific endeavour. That animal will not be generally available as and when somebody makes it in the three or four laboratories involved. But there you're making a product and I would guess again if you wanted to use it for research purposes it could be made available, but restrictively, then any products would go to the owner of the animal.

Garthoff: I really agree with that. Apparently we haven't differentiated between those animals we're using for research, no matter whether it's in industry or academia. On the other hand we have also transgenics that we're using for production of necessary remedies, and we mentioned yesterday α_1-antitrypsin. Of course, in that case I disagree with you. Those animals definitely will have some value and also the patent because what you've patented is more or less the process of getting that kind of a remedy. So, from that point of view those animals will definitely have their price and interest.

Waugaman: In the U.S. Academic laboratories are often the source of many of these new animals with special genomic characteristics that have immediate application in the research field. Insofar as most acedemic laboratories aren't set up to produce those kind of animals and provide them to research colleagues in other laboratories; that has to be done by a commercial venture. Most academic institutions in the U.S. look, as Dr. Rowland had said, at whether or not a patent is necessary before a patent is applied for. Is a patent necessary to allow a commercial partner to invest in providing that. If there's not the demand, why bother? There's no sense and no real necessity to invest institutional funds in applying for or maintaining a patent where there's no necessity to protect a market position.

Ganten: I'm getting through hot and cold showers at this meeting. This is the first meeting I've been to discussing patent and legal aspects and commercialization of products. We have one patent pending for transgenic animals and in the beginning we were very hesitant and then I thought it may not be so wrong if you earn a little bit of money for research with it and control what's being done with the animals. I was most impressed by the talk of Dr. Sahai with the possible consequences of patenting world-wide. I was not aware of that at all and the question I would really like to see discussed is whether patent law really affects what Dr. Sahai reports. Companies will control the world market anyhow probably whether they do it through patents or something else. Is this really a political problem on how we deal with the problems of this world or is it the patent law which in fact is instrumental in controlling this type of activity and flow of information and and seeds a.s.o. from one country to the other?

Hofschneider: We can turn to the plant situation and I would like to take up what Dr. Ganten said. If I understood you correctly and I repeat briefly what you said, it's that the patent laws of the North are more or less responsible for the lack of food in the South. For that reason the patent law has to be changed. Is that the correct repetition of your statement?

Sahai: No it's not. The first thing that I'd like to address myself to is what Dr. Rowland said, I didn't interrupt at that time because I didn't want to interrupt in that discussion. What he said was that it was India's problem to do what Queensland had done. That's not entirely true and the reason for that is that the 1983 Convention on Common Heritage is an international convention. It was very difficult even in the biodiversity convention to get a consensus – we've gone through four prep meetings – on accepting sovereignity over genetic resources. So, it is an international interdependent issue. We would now have to move into unilateral actions, but we'll also have to be at some level internationally interdependent and the whole question of sovereignity over resources is, linked to the question of access to resources which is the absolute crux of the Biodiversity Convention. So I'd like to repudiate what you said there about being a country-specific problem. To come back to Prof. Hofschneider, whether I think that the hunger in India is because of the patent laws of the North, no. I said that the situation will be exacerbated, I said that it is a completely ludicrous situation that a patent regime that has developed over a period of time interlinked to the development of the technological status of a country with which it is in harmony, – should just be taken out of its context and plunked into a developing country with *absolutely no* understanding of what the application of that international property will mean in that country; that it is not commensurate with the technological development of that country and would exacerbate the problems. It will deny us the right of self-reliant growth.

Garthoff: May I come back to the issue that you raised before – that is the question, as multinational companies exist, whether this more or less results in exploitation of, for example, plants or whatever exists in the third world countries. We can't solve the North-South difference just by changing patent laws, that's for sure. As you mentioned before, this is apparently a situation where things have been developing over centuries. I am not asking the same way in which you asked the national property for the plants that I would ask for the property, for example, of vegetables to the U.K. You also mentioned one issue, that is access to something. Of course, the access to something in a third world country has also to take into consideration that you have the chance to make use of that. That at the moment goes by the help of, for example, multinationals beside the national system that you have in India. We had the same situation in Eastern countries where we didn't have a patent law system at all, where this was more or less left to the state and federal planning of the socio-economic situation. If, for example, we want to have an anti-malaria substance and there was one for 2000 years around in China, this compound was never usable because the alkoloid in it was too labile. With the help of Western industry, by now the substance is being developed after 2000 years and made to a valuable anti-malaria substance which isn't in actual fact a death disease of the Western countries. So, what I'm saying is, yes of course, the multinationals are using resources out of third world countries as they do out of the Western countries, but it is also for the benefit of those that are in the third world countries. Of course, the whole system looks different than the way developing countries would like to see it. But this is the general North-South aspect that can't be changed by patent law.

Hofschneider: Do you agree that Bayer or Hoechst are number one and two in the seed and pharmaceutical industry?

Garthoff: This is always a question of ranking, there's always a list you put up and say O.K. there's number two in pharmaceuticals, number three etcetera depends on what kind of criteria you apply. I wouldn't comment on the ranking of that.

Sahai: To your statements about the relative contributions: Take this anti-malaria drug in China, if you recall my idea about what would be an equitable form of intellectual property protection. What I'm trying to say is that these factors will have to be taken into account which are not taken into account at present in any of the intellectual property regimes as they exist today. I am by no means pleading for an exclusive contract. All I'm saying is, and I've said it consistently, is that it will have to be one of the components that will have to enter into a property regime that would be equitable for us. The other thing is again something that I said already, does it all just come back to patent laws. Yes and no in the sense that some of it does come back to patent laws, I mentioned the point about compulsory licensing and treating importation as working of a patent. These are two clearly patent law issues. These are also two very clear issues with a heavy bias in favor of multinationals and industrial countries which are sought to being removed from the protection system that is being demanded. There's no clear reason why the clause of compulsory licensing and treating the importation as working should be removed from the patent law even if it has to be transposed per se because this would really take away every opportunity of technology dissemination. It is a question of patent law and finally, as I also said and I repeat again, all of us who are sitting in this room are bound to think beyond our narrow confines. We just cannot keep repeating that's not part of patent law, that's not part of science. All of us are opinion makers. We have to try to see broader issues for what they are and try to influence broader issues if we are convinced about them. I'm not satisfied only as a scientist. I think it would be very remiss for me just to lock my eyes into my lab and not look out the window. So, I plead again for a broader vision.

Izhar: I wasn't sure I understood what is the main message of Dr. Sahai. I think the solution for bridging the North-South gap is technology transfer in a very wide sense and this is being done to a certain extent. The patent issue is a very small issue and there is another thing which is reciprocity. If the developing countries want – and probably justified – to get some sort of royalties on their resources they will also have to pay for the outcome later. Another point that I'd like to stress about the resources: There's the primary source, the secondary source and there is perhaps the third source. Just take an example – wheat. The origin of the wheat is between Israel and Turkey. It flourished in Russia and from there it came to the new world, and the new world and its breeders contributed a lot to it. So, they also have to be compensated in some way. This is a very broad issue and a patent is a very small fraction of it.

Straus: Since I am possibly one of the acedemic protagonists for patenting I feel somehow addressed. First of all I feel quite comfortable with some remarks because for the first time I raised these questions some eight years ago in Mexico and in India. I was always of the opinion that the theory of the common heritage of mankind was somewhat imbalanced and not in place. And it's certainly true that the Rio Convention on Biodiversity has changed this point and I was quite pleased to see that even the United States from the outset accepted the new approach although did not sign at that time the convention. They did not question this principle. What bothers me is the fact that many people who feel they are opinion makers, tend to lose realistic sight as to potentials of Rio. As far as I can see, the only realistic approach is to safeguard an equitable share in genetic resources is to bring them under the umbrella of intellectual property because they are not crude oil. If in the melon case the United States knocked on your door if that gene were not in the gene bank in the United States but in Ethiopia, you could say, pay me one billion if you would estimate that. Maybe you will get it or not because maybe they will say, we will lose money if we pay. In my view the only somewhat promising way is to get the information stored in the genetic resources into the existing system. Maybe the system should and must be adjusted. Realistically, however, the chances for such changes are more than modes. You have the sovereignty of developing countries but the same applies for the others. They will not be willing to accept that. Maybe this is one point on which we could go a little more into depth in the round table discussion. And let me make a second remark: I never thought that local breeding wouldn't be necessary. Quite the contrary. I made this point also concerning the relationship between the United States and Europe. What we need for our agriculture is local breeding. What I was trying to say is that until now there were no intellectual property rights in most countries where the genetic diversity is located. Most of those developing countries are not even members of the Paris Convention. There was no single obstacle to use seed material from the developed world to adapt it to the local needs in those countries. There is no claim to say that there was an obstacle. You could have bought it and planted it and bred it. One should be fair enough to see the own weak points and the weak points of the others.

Sahai: I would like to address one or two of the things that Dr. Straus has mentioned. I don't think it is anybody's claim to claim a billion dollars and another erroneous assumption is that California would lose only one crop of melon. If you were not able to introduce resistance you would not be able to grow that crop again. It's not one crop. The other thing is if you've been dealing with the Biodiversity Convention, then I think the systems that have been proposed to deal with compensation for genes should be evidently clear to you. It's been nobody's claim so far that a billion dollars or ten thousand should be paid for an individual gene. There has been thought in systematic terms of some kind of collective compensation that could possibly be used for conservation mechanisms – the money that came out of farmers' rights or payment for genes. As for intellectual property rights in developing countries there is a very sound reason – and I went into considerable detail in my talk – as to why we don't have intellectual property rights because it's in the public domain. I think the answer lies in genuinely trying to address the needs of several possibly disparate interests.

Session 6

The Human Renin-Angiotensin-System in Transgenic Rats

D. Ganten and J. Wagner

Abstract

The renin-angiotensin-system is a major regulatory system of blood pressure and cardiovascular function. It has been implicated in the pathogenesis of hypertension in humans as well as in the development of cardiac and vascular hypertrophy. However, studies on the regulation of the genes of the renin-angiotensin-system and their impact on hypertension in humans are difficult to perform. We, therefore, established transgenic rats carrying the human renin or the human angiotensinogen genes. These rats showed tissue-specific expression of the transgenes and secreted the human proteins into the plasma. Due to the species-specificity of the human renin substrate reaction, human renin and angiotensinogen do not interact with the components of the renin-angiotensin-system of the rat. However, in transgenic rats carrying the human angiotensinogen, infusion of human renin elicits a hypertensive response, which can be blocked by a human renin specific inhibitor, Ro 42–5892. The "transfer" of the human renin substrate reaction to rats allows to test human specific enzyme kinetics in the rat, which are suitable models for cardiovascular research. These rats may therefore be test models for the development of human specific renin inhibitors and for studies on the interaction of the human RAS components in the pathogenesis of hypertension. These rats have been considered for patenting.

Introduction

The renin-angiotensin-system (RAS) is one of the best studied regulatory systems involved in control of cardiovascular function and volume homeostasis. Its effector peptide angiotensin II exerts a multiplicity of functions by raising peripheral resistance through vasoconstriction, enhancing renal sodium reabsorption, facilitation of catecholamine release from sympathetic nerve endings or stimulation of mineralcorticoid production in the adrenal gland (1, 2). These integrative functions focussing on control of body salt and water homeostasis as well as blood pressure have led to the hypothesis, that malfunction of this system may be related to the pathogenesis of hypertension (3, 4).

Since the forties it has been known that activation of the renin-angiotensin-system may indeed raise blood pressure as in the case of two kidney-one-clip hypertensive rats (5). Unilateral renal ischemia leads to secretion of renin from the "clipped" kidney, resulting in enhanced angiotensin II formation and raised blood pressure. This model is known as a high-renin hypertensive model, which is similar to human hypertensives with renal artery stenosis. The subdivision of hypertensive patients according to their plasma renin levels in high-, normal- or low renin-hypertension groups underscores the relevance, which has been attributed to this system in hypertension (6).

Due to the major role of the renin-angiotensin-system in blood pressure control, it has been considered as a candidate gene system for hypertension. Recently, the angiotensin-converting-enzyme gene has been found in a region of the rat chromosome which is linked to the hypertensive phenotype in the stroke-prone strain of spontaneously hypertensive rats (7).

Studies on the tissue-specific regulation of gene expression of the components of the RAS in humans are difficult to perform due to obvious ethical reasons and the small amounts of tissue available, e.g. from biopsies (8, 9). Since hypertension research is finally focussed on human hypertension and its treatment, the "transfer" of the human genes to the rat as a test model for cardiovascular research would be highly adavantageous to investigate the role of the human RAS components in hypertension. Therefore, transgenic rats may be helpful to fully understand the regulation of the human genes of the RAS in vivo. It offers the possibilitiy to study a hypertensive rat model, where blood pressure elevation is dependent on the interaction of the human RAS components.

A large body of data supports that the RAS may not only be involved in the development of hypertension, but also influence the process of cardiac and vascular hypertrophy (10, 11). These primarily physiological adaptive responses to elevated blood pressure progress in the course of hypertension to cardiac and renal failure. Blood pressure-independent, mitogenic and growthpromoting effects of angiotensin II on endothelial cells, fibroblasts and vascular smooth muscle cells, possibly via activation of oncogenes, are well known (12–14). These influences may mediate the deleterious maladaptations of the cardiovascular system.

Since coexpression of the components of the RAS has been demonstrated in the heart or the vasculature, angiotensin II in the cardiovascular system may not solely be derived from the plasma, but rather be produced by local, tissue-specific renin-angiotensin-systems and act on the surrounding tissue in a paracrine or autocrine fashion (15–17), for review). Expression of human renin and other components of the RAS in extrarenal tissues in transgenic rats may be a way to investigate the activity of tissue-specific human renin-angiotensin-systems on cardiovascular morphology and function.

Due to the widespread implications of the RAS in hypertension, an arsenal of pharmacological substances, which interrupt this system on every step of the cascade has been developed. The largest clinical experience has been obtained from angiotensin converting enzyme inhibitors (CEI), which now belong to the most effective and safest antihypertensive drugs. Additionally, angiotensin receptor antagonists and renin inhibitors are now clinically being studied for their use in antihypertensive treatment (18–20). Pharmacological studies using these substances, however, can not give a final answer on the role of the RAS and also have clinical drawbacks: Converting enzyme inhibitors, for example, are not specific for the RAS and cleave other vasoactive peptides, such as bradykinin. These may be responsible for some of the side effects of converting enzyme inhibitors such as painful cough (19). Angiotensin receptor antagonists raise plasma angiotensin II concentration and may increase binding of angiotensin II to other angiotensin receptor subtypes not being blocked by the particular antagonist and thus provoke unwanted side effects.

The rate limiting step of the enzymatic cascade of the RAS is the reaction of human renin with its substrate angiotensinogen (21). This highly specific reaction can be blocked by renin inhibitors. Potentially, blockade of this system at this early level could be therapeutically of advantage and may, at least theoretically, be superior to blockade of converting enzyme inhibition generally of in selected cases (22). However, the development of such renin inhibitors has been hampered by the species-specificty of the renin substrate reaction. Angiotensinogen is cleaved between the aminoacids at position 10 and 11. At position 13 a histidine is substituted for a tyrosine in human angiotensinogen, which is probably responsible for the species-specificity due to its charged group (23, 24). That precluded rats as test models, even though they are highly suitable for cardiovascular research and the testing of antihypertensive drugs. A large body of knowledge about the RAS could only be obtained by investigations performed in this animal species. Preclinical studies on renin inhibitors had to be restricted to primates such as marmosets or to humans in selected cases (20, 25). Transfer of the human renin substrate reaction to the rat would allow the use of rats as test models for the development of human-specific renin inhibitors. Therefore transgenic rats, carrying the human renin and angiotensiongen genes, are tools to address these questions and to measure the endocrinological, hemodynamic and functional responses to human renin inhibitors acutely or after chronic application.

Generation of Transgenic Rats

In transgenic experiments, foreign genes have been integrated in the genome of a number of species, such as mice, rats, goats and swine, but mice have been used most extensively in transgenic research. Several factors may account for this: there is a large database of mouse genetics and transgenic technology can easily be applied in this species. For some areas of research, however, the mouse may not be the most ideal species for transgenic studies. Its size puts limitations on the methodological arsenal which can be used for the detailed investigation of phenotypic changes induced by the expression of transgenes. One area of research where this is particularly true is cardiovascular biology. Although investigators have succeded in applying sophisticated techniques to the measurement of cardiovascular parameters in mice (26, 27) this is still considered an exception.

In addition, primary hypertension, for example, is virtually unknown in mice, whereas a great number of rat strains show this disorder (28). Experimental hypertension research, therefore, has largely been carried out using rats as a model and there is a great number of data available to serve as reference points for the measurement of cardiovascular parameters in this species. These considerations made it desirable to establish transgenic rats as experimental models in hypertension research. Transgenic animals are defined by the fact that new genetic material has been integrated in the genome by experimental methods other than breeding. Through incorporation into the germ cells, transgenic founder animals pass the transgene on to their offspring and a transgenic line can be established. This general method can also be applied for the establishment of transgenic rats (29), but there are several technical details, which have to be modified (30). One is superovulation, which tends to be more difficult to achieve in the rat. This problem can be solved, however, by adapting the procedure of Armstrong and Opavsky (31) using the application of follicle stimulating hormone (FSH) via osmotic minipump. By this approach the first transgenic rat line in cardiovascular research was established by Mullins, Peters and Ganten (32), who expressed the mouse *Ren-2* gene in rats.

A major application of transgenic techniques among others is the establishment of whole animal models, overexpressing a specific gene in the organism. In the following, we report the generation of transgenic rats harboring the human RAS components renin and angiotensinogen under control of their natural promoters. These transgenes in combination in one rat line can be a hypertensive rat model depending on the specific interaction of the human components of the renin-angiotensin system.

Transgenic Rats Carrying Genes of the Human Renin-Angiotensin-System

Transgenic Rats Harboring the Human Renin Gene

A construct containing the entire human renin gene was microinjected into pronuclei of fertilized rat oocytes from outbred Sprague-Dawley. The construct comprised a total length of 17.6 kb after stripping it from vector encoded sequences (33). It contained 10 exons, 9 introns and 3 kb of 5'-flanking sequence and 1,2 kb of 3'-flanking sequence. Two rat lines could be obtained which transmitted the transgene to their progeny (34).

These transgenic rats produced and secreted active human renin into their plasma, as has been determined by an immunoradiometric assay using monoclonal antibodies specific for human renin. The plasma levels of active human renin in one line was slightly less than in humans, whereas active human renin exceeded about twelve times the levels found in humans (1.5 ng ANGI/ml/h; (35)) in the other.

To examine, if the presence of the transgene interferes with the production of the endogenous host renin, plasma rat renin concentration was also determined, but no significant differences existed between transgenic rats and negative controls. Also, no altertions in other components of the RAS such as angiotensin I, II or angiotensinogen could be detected. Consequently, both rat strains were normotensive at about 133 ± 3 mmHg systolic as determined by tail plethysmography. These negative findings can be explained by the fact, that human renin does not interact with rat angiotensinogen except at very high doses. To investigate, how the human renin transgene responds to physiological stimuli of renin secretion, animals were sodium-depleted by initial intraperitoneal injection of furosemide (10 mg/kg) and placement on a low-sodium diet for three days. In plasma, active human renin levels increased by about 11-fold up to 58.3 ± 9 pg/ml, and rat renin slightly less by about 8 times. This indicates, that the transgene is regulated by sodium-depletion and does not interfere with rat renin production even under stimulated conditions.

Transgene expression was found to be highest in the kidney, but was also present in extrarenal tissues such as the lung or the gastrointestinal tract (34). On the cellular level, in-situ-hybridisation showed that human renin expression under basal conditions is confined to the juxtaglomerular apparatus. These findings are comparable to data obtained from the mouse, where human renin had also been expressed (36). Transgene expression in the mouse was confined to the vas afferens of the kidney, with expression of renin approximately 7-fold higher than in humans, but still about ten times less than mouse renin.

Transgenic Rats Harboring the Human Angiotensinogen Gene

The genomic human angiotensinogen gene construct, comprising 5 exons, 4 introns, 1.3 kb of 5'-flanking region and 2.4 kb of 3'-flanking region, was used to generate transgenic rats carrying the human angiotensinogen transgene (37). Four lines could be obtained which transmitted the transgene to their progeny. Human angiotensinogen was secreted into the plasma in all rats but the levels markedly varied between the lines ranging from 120 µg/ml up to 5 mg/ml. Thus, all four lines exceeded plasma angiotensiongen levels in humans which amounts to about 60 µg/ml (38).

Despite the high plasmatic levels, these animals were normotensive, indicating that human angiotensinogen did not interact with rat renin. Compared to transgene negative controls, rat angiotensinogen levels as well as angiotensin II levels were not significantly different in transgenics.

Transgene expression was highest at the appropriate sites in the liver, where it was expressed in the parenchyma. Transgene expression could also be detected about ten-fold lower in the kidney or the gastrointestinal tract. These findings are in contrast to a previously described transgenic mouse strain, carrying the human angiotensinogen gene. Here, human angiotensinogen was expressed as high in the kidney as in the liver in contrast to humans, where kidney angiotensinogen expression is low (39). The reasons for these differences in the quantity of tissue-specific expression are not completely understood, but may be due to the lack of negatively regulatory sequences beyond the 1.3 kb of 5'-flanking region, which may be necessary to suppress renal angiotensinogen expression.

Extrarenal Expression of Human Transgenes

Classically, the RAS has been viewed at as an endocrine system, where the active components are produced in the kidney (renin), liver (angiotensinogen) and lung (converting enzyme). Renin and angiotensinogen are then secreted into the plasma, where they produce blood-borne angiotensin II, that finally exerts its effects on the target organs. However, the blood pressure lowering effect of captopril in low-renin hypertension or in dialysis patients, who lost the ability to produce active renin from the kidneys, as well as the demonstration of plasma prorenin levels after nephrectomy, sometimes as high as in humans with normal kidneys, pointed towards an extrarenal production of renin (15, 40).

Renin expression could be detected in a number of other tissues such as the brain, the adrenals or the vasculature as well as in tissues which appear not to be primarily involved in cardiovascular regulation as in the gastrointestinal tract. Similarly, angiotensinogen and converting enzyme expression is widespread (41, 42). The coexistence of all the components of the RAS in one tissue has raised the hypothesis, that angiotensin II can be produced locally independent from the circulatory RAS, and act in an autocrine or paracrine fashion in the surrounding tissue (43). Such extrarenal, local RAS have been suggested for the brain, the adrenal, the cardiovascular and genitourinary system (15). The transgenic rats, carrying the mouse Ren-2 gene (see elsewhere in this book), develop a high blood pressure in presence of a low plasma but enhanced tissue-specific renin activity. This strengthens the contribution of the local RAS to the development of hypertension.

Human renin and angiotensinogen in transgenic rats have been shown to be expressed highest in the kidney or the liver. But expression is also demonstrable in extrarenal or extrahepatic tissues, respectively. Thus, human renin-mRNA is also present in adrenals, the brain, the gastrointestinal tract or the thyroid as well as the lung. In humans, renin expression in the gastrointestinal tract could be detected (44). Angiotensinogen can be found extrahepatically in the lung, the gastrointestinum, brain areas and the heart. The coexpression of the human transgenes in these organs supports the possibility of local angiotensin I production due to the interaction of the human proteins. Finally, angiotensin II can then be synthesized in presence of (non-species-specific) converting enzyme. The functional role of such human-dependent tissue-specific RAS remains to be established, but raises the question about the participation of tissue-specific RAS in human hypertension.

Species-specificity of the human renin substrate reaction – effect of the human renin inhibitor Ro 42-5892 –

Despite the high expression of the human transgenes, the rats remained normotensive as an indication of the species-specificity of the human renin substrate reaction. As demonstrated by the unaltered levels of angiotensin II in both human renin and angiotensinogen transgenic rats, neither rat renin reacted with human angiotensinogen nor did human renin and rat angiotensinogen. Blood pressure remained unaltered, when the human specific renin inhibitor Ro 42-5892 was given as a bolus injection at a dosage of 1.5 mg/kg body weight to sodium-depleted non-transgenic rats, whereas the angiotensin II receptor antagonist DuP 753 lowered blood pressure by about 20 mmHg. In contrast, 10^{-6} M Ro 42-5892 completely inhibited active human renin after incubation of plasma from sodium-depleted human renin transgenic rats, whereas rat renin remained unaffected.

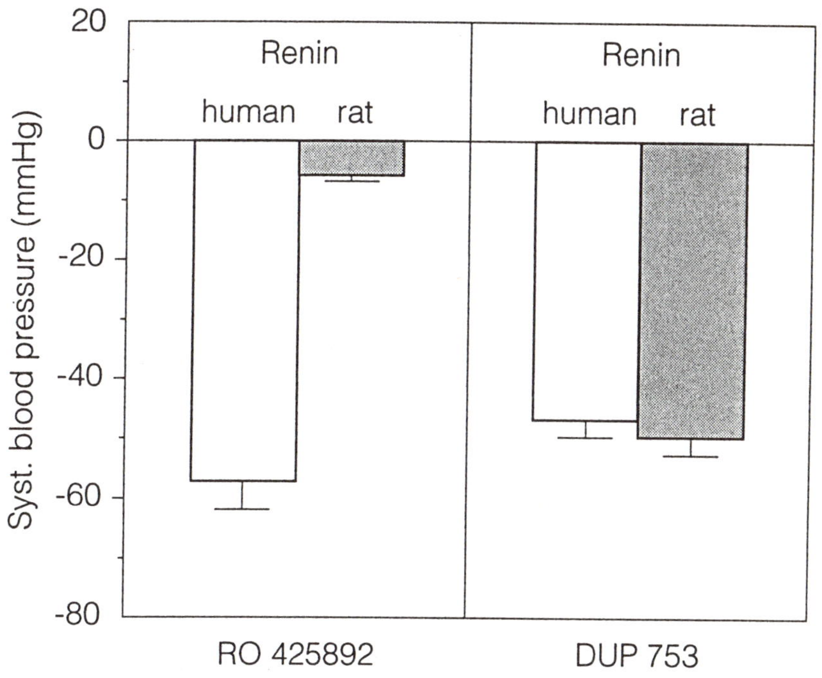

Fig. 1. Specificity of the human renin substrate reaction. Human renin infusion into transgenic rats carrying the human angiotensinogen gene elicits a hypertensive response, which can rapidly be reduced to normal by 1.5 mg/kg of the human renin specific renin inhibitor Ro 42-5892. Infusion of rat renin in equipressor doses also elevated blood pressure, but here, Ro 42-5892 remained without effect, whereas DuP 753 rapidly lowered blood pressure to pretreatment values. This indicates, that the blood pressure increase in response to human renin is due to the interaction of human renin with human angiotensinogen, which can specifically be blocked by the human renin inhibitor in vivo

These findings show, that human renin transgene does not add to blood pressure maintenance even after stimulation (data not shown).

The species-specificity of the human renin substrate reaction in the transgenic rats could further be demonstrated by injection of recombinant human renin into rats carrying the human angiotensinogen gene. Here, at a dosage of 5 µg ANG I/ml/h, blood pressure rapidly increased from 142 ± mmHg to 192 ± 8 mmHg. Addition of the human renin inhibitor Ro 42-5892 rapidly normalized the blood pressure to pretreatment values (Fig. 1). Human renin at this dosage did not elicit a hypertensive response in transgene negative controls, indicating that the blood pressure increase was due to the interaction of human renin with human angiotensinogen (Fig. 1).

Infusion of rat renin in equipressor doses raised blood pressure as well, but in this case, Ro 42-5892 remained without any effect, whereas DuP 753 (10 mg/kg) normalized blood pressure rapidly, supporting that angiotensin II formation in this case originated from the reaction of rat renin with rat angiotensinogen (Fig. 1).

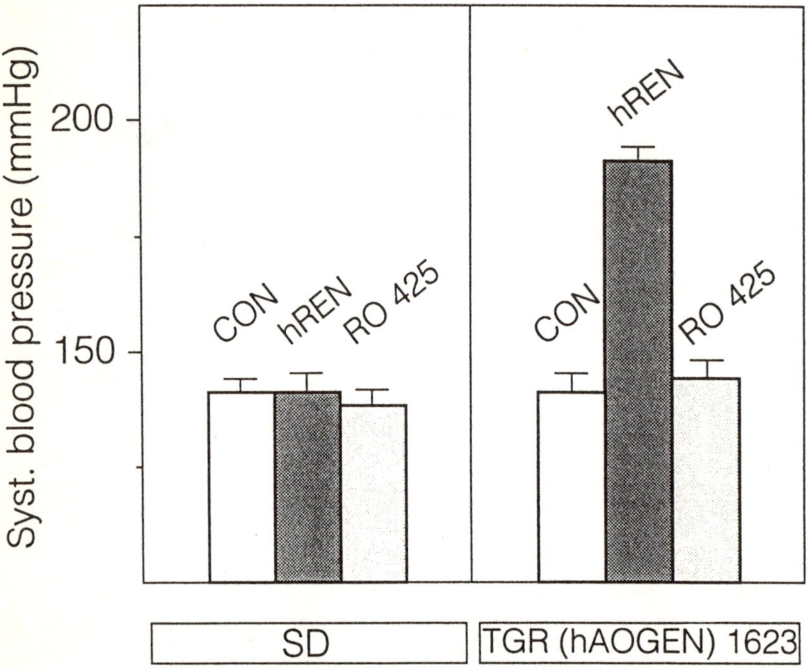

Fig. 2. In vivo specificity of the human renin-angiotensinogen interaction. Infusion of human renin into transgenic rats carrying the human angiotensinogen markedly raises blood pressure. Pretreatment of transgenic rats 3hrs prior to human renin infusion completely blocked blood pressure response. Plasma angiotensin II levels parallel the increase in blood pressure and are normal in presence of Ro 42-5892 (data not shown). In transgenic negative controls, no increase in blood pressure or angiotensin II is observed after infusion of human renin

Pretreatment of TGR (hAOGEN) with Ro 42-5892 completely blocked the hypertensive response as well as the increase in angiotensin II formation after injection of human renin (Fig. 2). The presence of DuP 753 (10 mg/kg i.v.) also prevented a blood pressure peak in presence of human renin.

The maintenance of the human specific renin substrate reaction in presence of chronic expression of the human renin or angiotensinogen transgenes allows the testing of renin inhibitors and their enzyme kinetics both in vitro and in vivo.

Conclusions

The general applicability of the transgenic rat model for basic research in cardiovascular biology and hypertension has now been established. This model offers advantages over the use of transgenic mice, particularly with respect to the characterization of cardiovascular parameters and pharmacological interventions, which are more readily feasible in this model. One of the first candidate systems thought to be involved in the pathogenesis of hypertension has been the RAS. The generation of rats expressing the human genes for renin and angiotensinogen have provided important models for the study not only of regulatory mechanisms of these genes, but also for the detailed investigation of possible pathological effects of its expression. Ultimately, the information obtained from the study of these transgenic animals should provide important insights into the regulation of the RAS in man. In addition, they also are experimental systems to study the pharmacological effects of specific substances interfering with the RAS in humans, such as renin inhibitors. Such experiments have the potential of providing information for the clinical use of these drugs, especially if they influence the specific sequelae of hypertension such as cardiac or vascular hypertrophy. Furthermore, the information that will be derived from future studies of those animals can be used to design experiments, which will further enhance our understanding of tissue specific gene expression and regulation using chimeric constructs composed of the human renin and angiotensinogen promoters and reporter genes. Experiments will also focus on the investigation of other candidate genes thought to be involved in human hypertension, such as the genes for angiotensin-converting-enzyme (ACE) and for the ANG II receptor.

Literature

1. Vecsei P, Hackenthal E, Ganten D (1978) The renin-angiotensin-aldosterone system. Past, present and future. Klin Wochenschr 56 (Suppl I):5–21
2. Klett C, Hellmann W, Hackenthal E, Ganten D (1990) Pathophysiologie des Renin-Angiotensin-Systems. Wiener Med Wschr 140:2–11
3. Bader M, Kreutz R. Wagner J, Zeh K, Böhm M, Paul M, Ganten D (1992) Primary hypertension and the renin angiotensin system: from the laboratory experiment to clinical relevance. In Colloque INSERM, Vol 218: Genetic Hypertension. John Libbey Eurotext, Montrouge, London, Paris. 359–370
4. Williams GH (1988) N Engl J Med 323:1517–1525
5. Goldblatt H, Lynch J, Hanzal RF, Summerville WW (1934) The production of persistent elevation of systolic blood pressure by means of renal ischemia. J Exp Med 59:347–379
6. Laragh JH, Sealey JE (1990) The renin-angiotensin-aldosterone-system in hypertensive disorders: A key to two forms of arteriolar vasoconstriction and a possible clue to the risk of vascular injury (heart attack and stroke) and prognosis. In: Laragh JG, Brenner BM (eds) Hypertension: Pathophysiology, Diagnosis and Treatment. Raven Press, New York
7. Hilbert P, Lindpaintner K, Beckmann JS, Serikawa T, Soubrier F, Dubay C, Cartwright P, DeGouyon B, Julier C, Takahasi S, Vincent M, Ganten D, Georges M, Lathrop GM (1991) Chromosomal mapping of two genetic loci associated with blood-pressure regulation in hereditary hypertensice rats. Nature 353:521–529
8. Brunneval P, Fournier JG, Soubrier F, Belair MF, DaSilva JL, Guettier C, Pinet F, Tardivel I, Corvol P, Bariety J, Camilleri J-P (1988) Detection and localization of renin messenger RNA in human pathologic tissues using in situ hybridization. Am J Pathol 131:320–330
9. Wagner J, Paul M, Ganten D, Ritz E (1991) Gene expression and quantification of components of the renin-angiotensin-system from human renal biopsies by the polymerase chain reaction. J Am Soc Nephrol 2:421 (Abstr)
10. Krieger JE, Dzau VJ (1991) Molecular biology of hypertension. Hypertension 18 (suppl I):I-3–I-17.
11. Morishita R, Higaki J, Miyazaki M, Ogihara T (1992) Possible role of the vascular renin-angiotensin system in hypertension and vascular hypertrophy. Hypertension 19 (suppl. II):II-62–II-67.
12. Schelling P, Fischer H, Ganten D (1991) Angiotensin and cell growth: a link to cardiovascular hypertrophy; J Hypertens 9:3–15
13. Paquet J-L, Baudouin-Legros M, Brunelle G, Meyer P (1990) Angiotensin II-induced proliferation of aortic myocytes in spontaneously hypertensive rats. J Hypertens 8:565–572
14. Taubman MB, Berk BC, Izumo S, Tsuda T, Alexander RW, Nadal-Ginard B (1989) Angiotensin induces c-fos mRNA in aortic smooth muscle: Role of Ca2+ mobilization and protein kinase C activation. J Biol Chem 264:526–530
15. Paul M, Bachmann J, Ganten D (1992) The tissue renin-angiotensin system in cardiovascular disease. Trends Cardiovasc Med 2:94–99
16. Lindpaintner K, Jin M, Wilhelm MJ, Suzuki F, Linz W, Schoelkens BA, lang RE, Unger T, Ganten D (1988) Intracardiac generation of angiotensin and its physiological role. Circulation 77:I-18–I-23
17. Baker KM, Booz GW, Dostal DE (1992) Cardiac actions of angiotensin II: role of an intracardiac renin-angiotensin system. Ann Rev Physiol 54:227–241

18. Rhaleb N-E, Rouissi N, Nantel F, D'Orleans-Juste P, Regoli D (1991) DuP 753 is a specific antagonist for the angiotensin receptor. Hypertensin 17:480–484
19. Poole MD, Postman DS (1991) Characterization of cough associated with converting enzyme inhibitors. Otolaryngol Head Neck Surgery 105:714–716
20. van den Meiracker AH, Admiraal PJJ, Man in't Veld AJ, Derkx FHM, Ritsema van Eck HJ, Mulder P, Van Brummelen P, Schalekamp MADH (1990) Prolonged blood pressure reduction by orally active renin inhibitor RO 42-5892 in essential hypertension. Br Med J 301:205–210
21. Sealey JE, Laragh JG (1990) in: Laragh JH, Brenner BM (eds) Hypertension Pathophysiology, Diagnoses and Management. Raven Press, New York, pp 1287–1318
22. Corvol P, Chauveau D, Jeunemaitre X, Menard J (1990) Human renin inhibitor peptides. Hypertension 16:1–11
23. Evans DB, Cornette JC, Sawyer TK, Staples DJ, De Vaux AE, Sharma SK (1990) Substrate specificity and inhibitor structure activity relationships of recombinante human renin: implications in the in vivo evaluation of renin inhibitors. Biotechnol Appl Biochem 12:161–175
24. Tewksbury DA, Dart RA, Travis J (1981) The amino terminal amino acid sequence of human angiotensinogen. J Biochem Biophys Res Commun 99:1311–1315
25. Ii Y, Murakami E, Hiwada K (1991) Effect of renin inhibitor, ES-8891, on renal renin secretion and storage in the marmoset: comparison with captopril. J Hypertens 9:1119–1125
26. Mockrin SC, Dzau VJ, Gross KW, Horan MJ (1991) Transgenic animals: new approaches to hypertension research. Hypertension 17:394–399
27. Field LJ (1991) Cardiovascular research in transgenic animals. Trends Cardiovasc Med 1:141–146
28. Ganten D, Lindpaintner K, Ganten U, Peters J, Zimmermann F, Bader M, Mullins J (1991) Transgenic rats: New animal models in hypertension research. Hypertension 17:843–855
29. Wagner J, Zeh K, Paul M (1992) Transgenic rats in hypertension research. J Hypertens 10:601–605
30. Mullins JJ, Ganten D (1990) Transgenic animals: new approaches to hypertension research. J Hypertens 8 (suppl 7):S35–S37
31. Armstrong DT, Opavsky MA (1988) Superovulation of immature rats by continuous infusion of FSH. Biol Reprod 39:511–518
32. Mullins JJ, Peters J, Ganten D (1990) Fulminant hypertension in transgenic rats harbouring the mouse Ren-2 gene. Nature 344:541–544
33. Fukamizu A, Seo MS, Hatae T, Yokoyama M, Nomura T, Katsuki M, Murakami K (1989) Tissue-specific expression of the human renin gene in transgenic mice. Biochem Biophys Res Commun 165:826–832
34. Ganten D, Wagner J, Zeh K, Bader M, Michel J-B, Paul M, Zimmermann F, Ruf P, Hilgenfeldt U, Ganten U, Kaling M, Bachmann S, Fukamizu A, Mullins JJ, Murakami K (1992) Species specificity of renin kinetics in transgenic rats harboring the human renin and angiotensinogen genes. Proc Natl Acad Sci USA vol 89:7806–7810
35. Menard J, Guyenne T-T, Corvol P, Pau B, Simon D, Roncucci R (1985) Direct immunometric assay of active renin in human plasma. J Hypertens 3 (suppl 3): S275–S278
36. Fukamizu A, Hatae T, Kon Y, Sugimura M, Hasegawa T, Yokoyama M, Nomura T, Katsuki M, Murakami K (1991) Human renin in transgenic mouse kidney is localized to juxtaglomerular cells. Biochem J 278:601–603

37. Fukamizu A, Takahashi S, Seo MS, Tada M, Tanimoto K, Uehara S, Murakami K (1990) Structure and expression of the human angiotensinogen gene. J Biol Chem 265:7576–7582
38. Gardes J, Bouhnik J, Clauser E, Corvol P, Ménard J (1982) Role of angiotensinogen in blood pressure homeostasis. Hypertension 4:185–189
39. Takahashi S, Fukamizu A, Hasegawa T, Yokoyama M, Nomura T, Katsuki M, Murakami K (1991) Expression of the human angiotensinogen gene in transgenic mice and transfected cells. Biochem Biophys Res Commun 180:1103–1109
40. Sealey JE, White RP, Laragh JH, Rubin AL (1977) Plasma prorenin and renin in anephric patients. Circ Res 41 (suppl II):17–21
41. Unger T, Gohlke P, Paul M, Rettig R (1991) Tissue renin-angiotensin systems: fact or fiction; J Cardiovasc Pharmacol 18 (suppl 2):S20–S25
42. Dzau VJ, Pratt RE (1986) Renin-angiotensin system: biology, physiology, and pharmacology. The Heart and Cardiovasc Syst 2:1631–1662
43. Dzau VJ (1987) Implications of local angiotensin production in cardiovascular physiology and pharmacology. Am J Cardiol 59:59A–65A
44. Seo MS, Fukamizu A, Saito T, Murakami K (1991) Identification of a previously unrecognized production site of human renin. Biochim Biophys Acta 1129:87–89

Which Rules of Patenting are Acceptable from the Point of View of Industry?

B. Yorke

Firstly, I would like to express appreciation to Prof. Vogel for taking the initiative to organise this Workshop. The area we are dealing with is a complicated one and fraught with misunderstanding. A Workshop such as the present gives the opportunity to understand the real reasons behind the point of view of others and to remove misunderstandings.

I am supposed to give the point of view of industry. However, there are many sectors of industry and they have different priorities and requirements. I may, however, be able to give a reasonable representation of the point of view of the research-based chemical industry. I am assisted in this by being head of the Patents and Trademark Division of a company which comprises the following six divisional companies: Pharmaceuticals, Seeds, Agrochemicals, Chemicals, Construction Chemicals and Environment, and Nutrition. Nearly all these divisional companies are involved in Biotechnology, although some more than others. I am also head of the Industrial Property Working Groups of the European pharmaceutical organisation (EFPIA) and the European chemicals organisation (CEFIC). On the other hand, I must confess that when I read chemistry at Oxford, we were studying the original Watson Crick publication on the double helix and merely theorising about what might result therefrom!

While I shall use my outline that you have in the documentation as a framework, I do not want to give a prepared talk based thereon but rather comment on some of the points raised by previous speakers. You will forgive me if this leads to a certain incoherence of development.

To be patentable, an invention must be new, involve an inventive step and be susceptible of industrial application (European requirements but similar in other countries, e.g. USA and Japan). It should not be contrary to "ordre public" or morality.

To avoid repetition I wish to concentrate on the last sentence. Although it is in the Law, I am very unhappy with the idea of a Patent Office Examiner, e.g. a European Patent Office Examiner, refusing an invention because it is contrary to "ordre public" or morality. Firstly, Patent Office Examiners are not qualified to take such decisions. It is, indeed, difficult to know who would be qualified but the last responsibility will be with judges. As, however, a previous speaker has said, how do you resolve for such issues the fact that Sicily and Copenhagen are both part of the EC but the Patent Office has to apply a single standard. Even if Patent Office Examiners were qualified, the process the European Patent Office seems to be adopting, namely weighing up the benefits of the invention for mankind against negative points such as the degree of suffering of an animal, is problematic. For example, at the timepoint of Patent Office examination, the true benefits of the invention can often not be properly assessed. One must also take into consideration that the present European system of patents (but not the US system) has departed from the original philosophy, namely that in return for disclosing his invention to the public rather than keeping it secret, the inventor should be rewarded with a patent. Nowadays in Europe the patent application is published long before the Patent Office examination takes place and the inventor can well end up with nothing although he has already disclosed his invention. When associated with a somewhat arbitrary and not definable ground of rejection, there seems to be an element of real injustice here. Finally, it should not be forgotten that if the Patent Office rejects the application the applicant has lost all his rights irrevocably. On the other hand, if the patent is granted it can always be attacked, e.g. in the Courts, afterwards.

For all the above reasons I believe that Patent Offices should not refuse inventions as being contrary to "ordre public" or morality except in what would generally be regarded as the very crassest cases. Otherwise, the issue should be left to the Courts if and when a problem arises.

A patent does not give the right to do anything, only to prevent commercial acitivities of imitatiors for a limited period of time.

This point cannot be emphasised enough. The effect of a patent is misunderstood by a large segment of the public. For example, if you have a patent to a new product, such as a pharmaceutical or agrochemical, this does not mean that you can market that pharmaceutical or agrochemical. One must first go through the massive task of obtaining approval from the regulatory authorities. Or there can be other laws against the use of your invention. Thus, the opponents of atomic energy in various countries did not need to attack patents in this area to reach their aims. If no new atomic energy plants are allowed to come into operation, the fact that you have a patent on an improvement in such a plant is essentially irrelevant. Another illustration of the fact that a patent does not give the right to do anything except to prevent others using your invention has been a subject of some concern to previous speakers. Thus there can be an earlier broad patent which dominates the invention of a later narrower patent. For example, a patent could cover a group of pharmaceuticals and a later patent could be issued to a particular pharmaceutical falling within the group but not specifically disclosed in the earlier patent. The later patentee needs a license from the former patentee to commercialise his pharmaceutical, whereby the earlier patentee would also need a license from the later patentee to commercialise that pharmaceutical. (Interestingly enough, such concerns were expressed when Germany introduced product protection for chemicals in 1968 – here one was concerned what would happen when someone developed a valuable use for a chemical already protected per se.) However the patent system has been around for several hundreds of years and has become a sophisticated instrument quite capable of dealing with such situations. The answer is voluntary licensing or cross-licensing, and this is what in practice happens. I have been in industrial patent practice for more than 25 years and I have never known such a situation which was not resolved in this way where truly different produces (or different uses) were involved. Thus our essential interest with our pharmaceutical patents is to prevent a pirate riding on our registration data and simply imitating our particular product.

Pure research does not constitute patent infringement, there must be commercial intention.

While the law can vary in detail from country to country, the above expresses what I believe the basic law to be in Europe, certainly as it manifests itself in practice. Nonetheless certain participants in this Workshop, e.g. researchers in universities, have expressed concerns that patents in this field will interfere with their research. Again, similar concerns were expressed when product protection for chemicals was introduced in Germany, but this did not lead to any problem in practice. I cannot conceive that a company such as mine would sue a research worker in a German university. What we are interested in doing is taking patent action against commercial competitors, e.g. against a generic manufacturer carrying out bioequivalency trials with the aim of introducing one of our pharmaceuticals within the patent term. The patent system is a mature one. It does not tolerate abuses or unjust anomalies. It is self-regulating in this. To take the matter a step further, some participants have expressed concern and rejection of certain terms DuPont wish to impose in connection with the patented onco mouse. To the extent that such terms may be inequitable, I am sure that operation of the patent system/commercial realities will change the situation. (N.B. Some time after the Workshop, it was announced that DuPont were changing the terms in connection with their onco mouse.) The situation in the United States is somewhat different, both as regards the law and regarding the activities of universities which are definitely becoming involved in commercial situations. It may well be that European universities will become increasingly involved with research having a commercial intention rather than pure research. This is then a different game with different rules.

While two days ago I would have been of a different opinion, participation at this Workshop and hearing the widespread concerns expressed lead me to consider that some clarification of the relationship between research and patent infringement may, indeed, be worthy of consideration.

Patenting of an invention automatically brings publication of the invention

I included the above statement in my outline as a result of previous discussions with opponents of biotech patents, indeed of modern biotechnology itself, with the aim of making the point that patents at least bring early transparency as to the direction research is taking. At the present Workshop, however, it is rather the reserve point which has arisen. Thus there have been strong plaidoyers for introduction of a grace period during which an invention may be published without this publication invalidating a subsequent patent application. Indeed such grace periods were provided for in the law of several states before the European Patent

Convention came into force. Industry has traditionally been against grace periods because of the legal uncertainty they produce. In the absence of a grace period, one can normally assume that if there has been a prior publication any subsequent patent is invalid. With a grace period situation, the effect of a publication can often be very unclear. Was this one by the inventor which is covered by the grace period? Is any other publication simply a republication of the original inventor's publication or was it a genuine non-connected third-party publication which would invalidate? Etc., etc. It happens, however, that the question of grace periods is already being discussed in another context, namely the international patent law harmonisation negotiations under the auspices of WIPO (World Industrial Property Organisation). A major issue here is the United States "first-to-invent" system for patents, rather than "first-to-file" as used in the rest of the world. The US "first-to-invent" system is particularly invidious seen from non-US eyes since it is effectively a "first-to-invent in the USA" system, and therefore discriminates against non-US inventors. The US system also leads to the notorious "interference proceedings" which are long, massively complicated and exorbitantly expensive. Furthermore, the outcome is very difficult to predict for several years and such uncertainty is anathema to industry, particularly when massive investment projects are involved. The Americans have shown some readiness to move to a "first-to-file" system but they are imposing certain conditions for this, one of which notably is the introduction of a grace period in those countries which do not have it, in particular Europe. (You will recall that the famous Cohen-Boyer patent was patentable in the United States because of the grace period there but unpatentable outside the United States because it had been published too early.)

Accordingly, the question of introduction (or re-introduction) of a grace period in Europe will probably be solved in connection with the patent law harmonisation negotiations. I personally would certainly support its introduction in the context of the Americans changing their "first-to-invent" system and I believe the majority of at least the chemical industry would do the same.

Biotechnology has the potential to bring massive benefits to the whole of mankind, not least in health and agriculture. Patents are essential if this is to be achieved

It is well established that today it takes some ten years of massively complicated research and development plus an investment of some DM 400,000 million in order to be able to bring a new pharmaceutical to the market. Moreover, the chances of success are low. Thus some 10,000 compounds are synthesised for each compound that reaches the market. Nobody would take such risks and invest so much time, skill and money if his product could then be immediately imitated by companies who have taken none of the risk and simply appropriate the pharmacological results, toxicological results, clinical results etc. Analogous considerations apply to biotechnological inventions.

It is for such reasons that patents are essential. In the absence of the limited period of exclusivity that patents provide, we would not have had the improvements in treating e.g. infections and other diseases in the past, nor will we achieve the improvements in treating terrible diseases such as cancer that are so desperately needed. Here we are talking about a free-market democratic society, rather than a totalitarian society whether of the left or the right politically. Until recently, one might have thought that socialism could work. Means would be invested for the good of the whole people. However, in practice the matter looks very different, as we know. I have spoken with Soviet factory managers. In the Soviet Union, the classical patent system effectively did not apply. Any patented inventions could be used by the State. So what did a factory manager do? He was not prepared to invest time and energy and money in making improvements, which make his results look less good, when he knew that the factory manager across the road would be able to immediately profit from these improvements without the investment.

Similar considerations apply to the transfer of technology to, and investment in, developing countries. In the absence of strong patent protection in these countries, the owner of the technology is hesitant about transfer, for there is nothing to prevent it "leaking" to his competitors.

There is also the myth that patents are only of real interest to the "big bad" multi-national companies. This is not the case. To the contrary, they are more significant to the small and middle-sized companies. If these make a useful invention, it is the existence of patents that enables them to profit therefrom. In the absence of patents, they would be swamped by the superior marketing capability of the large companies.

The NIH patenting of human gene fragments is an unfortunate "red herring" in the debate

I do not wish to spend much time on this. My attitude is summarised above. There has been massive public debate on this matter, a large portion of which is misinformed. If an invention is novel, unobvious and useful, which patent law requires, then I believe patents should be granted. It could well be that the NIH activities will fall on these patent law criteria. In any event, patent law is mature enough to be able to deal with this situation and, if gene fragments are patented, to make sure that this does not prejudice legitimate activities. However, such red herrings always have a unfortunate effect when they achieve a large measure of publicity. For example, the European Commission initiative for a Directive concerning the patenting of biotechnological inventions already, because of the NIH affair, contains superfluous wording which probably will have negative effects in the future.

The patenting of living organisms does not take anything away from farmers, developing countries etc. It will provide them with new alternatives

One reads many statements in this area which to a patent expert are incomprehensible. For example that patenting will stop farmers doing in developing countries what they have been doing for hundreds of years. How can this be the case when a patent has to relate to something new? There will undoubtedly be developments of massive benefit to developing countries, for example new plants that can exist in very poor dry soil. But these developments do not take anything away, they merely provide new alternatives. For this reason, I do not really understand why patenting will allegedly reduce biodiversity, etc.

I am sure everyone here has genuine concern for the problems of the developing countries. But is attacking the patent system an answer to these problems? I am personally convinced that it is not. Strong patents would not harm developing countries and they have the potentiality to do good. It is no coincidence that the United States in its original constitution provided for strong patents. At the time, it was certainly a developing country and had just ceased to be a colonial country. Due to, mainly, historical accident, various countries which not so long ago were still developing countries or still are have long had strong patent systems. Example are Hong Kong, Singapore etc., etc. Have these countries suffered? Even if we move to the developed world, the experience is the same. For example, due to an edict of Mussolini, Italy hat no protection whatsoever for pharmaceuticals until 1978. At this time the Supreme Court decided that strong protection for pharmaceuticals was in fact in the interests of the general public in Italy. We heard all the usual clichés. Drug prices would explode, Italian companies would go bankrupt, research and development would cease etc. But what are the facts? Prices did not explode, they increased less than the general inflation. Research and development in the pharmaceutical area, employment, all increased massively etc. It seems to me that the main problem we have in this area is theoretical propaganda coming from, for example, economists in UNCTAD in the 60's. A great experiment with such an approach came from India. But what happened in India when some 20 years ago they effectively got rid of pharmaceutical and agrochemical patents? Did the Indian public at large get better access to the essential drugs that they desperately need? No! Indeed this is another red herring because virtually no drugs on the World Health Organisation's list of essential drugs, precisely those drugs which are important for the poor Indian inhabitants would have been under patent protection even if India had had the strongest patent protection of the Western world. But imitator companies in India and their owners did profit from this situation and there is much cynical, idealistic sounding propaganda, today in India and in various Latin American countries whose real purpose is to protect the narrow interests of these business men in those countries. I am convinced that the quicker one mores away from centrally planned socialist societies in the developing countries, the greater the chances that these countries will have to progress.

The ethical debate over research into constituents of the human body, genetically changed animals, plants etc. is taking place in the wrong context. Patents are secondary; it is not the borders of patentability over which society has to decide, but the borders of research.

The onco-mouse as illustration. The patenting takes nothing away from the mouse, changes it in no way. The mouse was already there before the patenting. Whether patented or not, the mouse remains the same. The essence of the onco-mouse is to help produce medicaments against cancer.

There will never be a patented human being

Professor Beier has quoted Baroness Mary Warnock "I can never understand how patenting does any harm to animals". Professor Zimmerli has given an erudite analysis leading to the conclusion that there are no problems from the ethical point of view regarding *the patenting* of biotechnological inventions. Accordingly, I do not need to pursue the ethical considerations regarding patenting in further detail here.

Nonetheless, the patenting of the onco-mouse in Europe has led to a concerted action by a wide range of organisations to oppose the European patent and to attack patents on living organisms generally. While I respect, but do not share, the opinion of religious groups who fear one is interfering with God's creation – but what is surgery? – my discussions with many of the opponents of such patents indicate that a large number of them are against biotechnology as such and indeed against free enterprise and industry generally. I, personally, would not go to the barricades to ensure that inventions such as the onco-mouse remain patentable. My company has made similar inventions, e.g. an Alzheimer mouse, that we have not patented. In my view, the true role of such patents only comes into play if one is in the business of selling such animals which my company is not. However, other companies are and, more important, a principle is at stake. The opponents wish to ban the patenting of living organisms, e.g. animals, completely and here one must go to the barricades because potential inventions of immense benefit to mankind are involved. An example concerns transplantation. Many of us are aware of the desperate plight of persons, often children, waiting for a suitable organ for transplantation. On the other hand, we have read horror stories of the origin of some of these organs. A transgenic pig which could provide organs which would not be rejected is for me an animal which should and must be patented. Similar considerations apply in the area of modified or isolated genes, human cells etc. in connection with, e.g., somatic gene therapy.

> *The rules of patenting for bio- and genetechnology should be the same as those for other areas of technology. Only so can the full benefit for mankind be realised*

I still support the above statement 100 %! However, Dr. Grunwald, Prof. Beier and other speakers have made various proposals, some of which I have referred to hereinbefore, which require comment.

The (re)introduction of a "grace period" into European law merits careful consideration and, indeed, will probably take place in connection with WIPO efforts to harmonize international patent law.

Some clarification of the law concerning research activities and patent infringement would probably be desirable.

There has also been some discussion of discoveries, e.g. the isolation of substances occurring in nature, with the suggestion that such should be claimed narrowly. I believe, in fact, that present patent law is fully capable of dealing with such situations equitably. The provision, e.g., of a pure protein such that it can be used as a medicament can certainly be a patentable invention, subject to the normal requirements of patentability despite the fact that minor amounts of this protein may exist, in a complicated environment, in the human body. What the patentee claims must of course be limited to, e.g. his pure product, distinguishing from what occurs in nature.

Finally, there have been suggestions that in the field under discussion product protection should not be provided and/or that compulsory licensing should specially be available. The research-based industry strongly rejects both of these suggestions as being unnecessary and harmful.

Quite generally, the patent system has served society well. It is ill-advised to make detailed changes at a particular moment in time due to perceptions (or political pressure) then prevailing. Patent law is generally sophisticated enough to deal with the issues of new technologies in an equitable way. Once a detailed change has been made it is difficult and requires a long time to reverse and in the meantime paralysis occurs. Swiss law recently introduced a ban on germ line therapy into the Constitution. I wonder whether this will prove to have been wise.

Extreme care, objectivity and rationality is required in the emotional atmosphere presently prevailing (which is often based on misunderstandings). Otherwise we will throw out the baby with the bathwater!

In response to Prof. Beier, research-based industry would not be able to support any derogation from product protection nor the introduction of compulsory licenses. Such proposals are reminiscent of those that were made when Germany introduced product protection for ordinary chemicals and arose out of similar fears. These fears proved groundless for product claims for chemicals and they will prove groundless in the present area. There is no need for such measures and they would prove very deleterious.

Patentability of Human Genes and Living Organisms: Principles of a Possible International Understanding

F.-K. Beier and R. Moufang

1. Introduction

1.1. We have been asked by the organizers to suggest possible principles of an international understanding concerning the patentability of living organisms and human genes. After the extensive discussions of the last two days, this is really no easy task. We are confronted here with several, highly controversial issues, as it has once more become obvious during the present workshop. The debate on the challenges to patent law posed by biotechnology has now been going on for some time. The famous Craig Venter patent application of the NIH claiming inter alia a great number of human cDNA-sequences of unknown function has added new fuel to this debate.[1] Critical comments raised by the scientific community as well as industry circles found their way into the public opinion. Even an international agreement forbidding the patenting of human gene sequences of unknown utility has been proposed.[2] However, the growing awareness of the general public concerning the patenting of biological entities stands, at times, in strange contrast to the frequent lack of interest of politicians, top industry managers as well as scientists, towards patent law.[3] This is the main reason for many misconceptions about basic features of this very special field of law.

1.2. In order to promote technical, economic and social progress, society once invented the patent system and has for centuries made use of this simple, inexpensive but rather effective legal instrument for the protection of new inventions. Despite the changes that have taken place and the great differences in the economic and social parameters, the main objectives of patent protection have basically remained unchanged since the first inventors' privileges were granted at the end of the middle ages. Exclusive patent rights are granted for a limited time to recognize the intellectual property of the inventor in the practical application of his ideas, to reward him for his useful services to society, to encourage invention,

---

1 Cf. Eisenberg, Genes, Patents and Product Development, 257 Science 903 ff. (1992); Kiley, Patents on Complementary DNA Fragments?, 257 Science 915 ff. (1992).

2 Cf. Howard, Patenting Complementary DNA, 256 Science 11 (1992); Roberts, Gene Patents – Scientists Voice Their Opposition, 256 Science 1273 f. (1992).

3 Cf. Beier, Gewerblicher Rechtsschutz, Soziale Marktwirtschaft und Europäischer Binnenmarkt, GRUR 1992, 228, 234 f.

investment and innovation, particularly by private industry, and, finally, to promote the dissemination of scientific and technical knowledge.[4]

1.3. It has to be recognized from the very outset that patent law as a legal instrument conceived with the aim of promoting technlogy can only with great difficulty act as an efficient safeguard against the abuses of dangers of new technologies. This basic insight was recalled in a very recent decision of the European Patent Office, in which the Opposition Division rejected the objections of Greenpeace against a patent[5] granted on a herbicide-resistant plant and stated:

> "The effect of a patent is thus to secure the rights of its owner to the invention. However, the patent does not confer a positive right to use an invention; exploitation of the patent is always subject to regulation by governmental agencies where appropriate (for example the patenting of a potential novel medicament has no bearing on whether it will be approved for administration to patients or not, nor does the grant of a patent for a plant imply that it will be approved for field trials, let alone for large-scale industrial use. . . . (T)he grant of a patent is completely independent of the existence of possible restrictions on the use of the patented invention; moreover, denying a patent does not in itself prevent an invention being exploited."[6]

1.4. In order to reach a common understanding of the present and future role of patent law in the area of biological innovations, it is necessary to analyze and discuss the major legal policy arguments that have appeared in the debate up to now. The Max-Planck-Institute and its researchers were fortunate enough to be able to take part in these discussions from their very beginnings when the OECD – as the first international organization to address the problem – commissioned a thorough analysis of the new questions created by modern biotechnology in the field of patent law.[7] This participation in the work of international organizations, such as WIPO,[8] UPOV,[9] UNCTAD,[10] UNIDO[11] and the EC, has led our Institute to

---

[4] For details on these patent theories cf. Machlup, An Economic Review of the Patent System, Study No. 15, Subcommittee on Patents, Trade Marks, and Copyright of the Committee on the Judiciary, US Senate, 85th Congress Second Session, Washington 1958 = Die wirtschaftlichen Grundlagen des Patentrechts, GRUR Int. 1961, 373 ff., 473 ff., 524 ff.

[5] European patent no. 242 236.

[6] Decision on application no. 87400141.5, p. 9 (sub 3.3).

[7] Cf. Beier/Crespi/Straus, Patent Protection in Biotechnology: An International Review, OECD Paris, 1984.

[8] Cf. Straus, Gewerblicher Rechtsschutz für biotechnologische Erfindungen, Cologne etc. 1987.

[9] v. Pechmann/Straus, Die Diplomatische Konferenz zur Revision des Internationalen Übereinkommens zum Schutz von Pflanzenzüchtungen, GRUR Int. 1991, 507 ff.

[10] Straus, Biotechnologia i jej miedzynaradowe prawne i ekenomiczne implikacje, 56 Prace z wynalazczosci i ochrony wlasnosci intelektualnej 11 ff. (1990).

play a certain role in promoting the awareness of the new challenges and in supporting initiatives to improve the legal situation.[12]

We will structure our contribution in the following way: First, we will address the question of whether and to which extent living organisms and other biological entities should be patentable subject-matter. Second we will deal with the problem as to how patent law should handle inventions still in the early stages of the innovative process. Third, possible policy options with respect to the interface of publicly funded research and patent law will be examined.

2. Living Organisms and Other Biological Entities as Patentable Subject-Matter?

2.1 Patentable Subject-Matter in General

Patent lawyers are used to distinguish between the question of whether a given innovation constitutes patentable subject-matter and the issue of whether it meets the patentability requirements, i.e. novelty, inventive step and industrial applicability. What is considered to be patentable subject-matter defines the metes and bounds of the area where innovative activity may be stimulated and rewarded. De lege lata, two conditions have to be fulfilled in this respect:

First, the given innovation ought to be an invention. This term constitutes the basis concept of patent law and has undergone considerable evolution in patent law history. With good reasons, the legislators of the EPC and of the German patent statute refrained from putting the concept of invention into the straight-jacket of a legal definition. In German case law, invention has been defined as a teaching to methodically utilize controllable natural forces to achieve a causal, perceivable result or – shorter – as "as technical teaching".[13]

Second, the given innovation must not be excluded by explicit provisions. In principle, the legislator can restrict the scope of what constitutes patentable subject-matter, although in the present international environment he is bound to a certain degree by existing treaty obligations.[14] At any rate, the legislator is well advised to proceed very cautiously when drafting exclusionary provisions. Patent systems are based on the premise that protection should in principle be available for all inventions. This is even anchored in the basic norm of both the European

---

11 Cf. Straus/Moufang, Legal Aspects of Acquiring, Holding and Utilizing Patents with Reference to the Activities of the International Centre for Genetic Engineering and Biotechnology (ICGEB), UNIDO-Doc. ICGEB/Prep. Comm./14/3 Add. 1.

12 Cf. Beier, Gewerblicher Rechtsschutz für moderne biotechnologische Verfahren und Produkte, GRUR Int. 1990, 219 ff.; Moufang, Genetische Erfindungen im gewerblichen Rechtsschutz, Cologne etc. 1988; Neumeier, Sortenschutz und/oder Patentschutz für Pflanzenzüchtungen, Cologne etc. 1990.

13 Cf. BGH GRUR 1969, 672 – "Rote Taube" = 1 IIC 136 (1970).

14 For instance, the Strasbourg Convention on the Unification of Certain Points of Substantive Law on Patents for Inventions of 27 November, 1963.

Patent Convention and the German patent law[15] which stipulates that patents shall be granted for inventions that meet certain requirements. Exemptions for specific fields of technical innovation constitute the ultima ratio of patent law and compelling reasons are needed to justify them. We propose to discuss the issue of patentable subject-matter under the following headings: Biological innovations and living organisms in general, plant inventions, animal inventions and inventions concerning the human body and body parts.

2.2. Biological Innovations in General

What could lead one to the assumption that biological innovations in general should not be encompassed in the field of patent law? There are some possible lines of argument, but none of them appear to be convincing.

One might first assert that patent law was initially conceived for innovative activity in the field of physics and chemistry, not for biology, and that for this historical reason, its scope of application should remain as it was. Such purely retrospective argumentation is utterly devoid of merit. When science and technology develop and change, it must generally be possible to bring their newer facets under the protection of patent law. This has been stated in very clear terms by the German Federal Supreme Court in its famous "Red Dove" decision already some 25 years ago[16] and still holds true. A less formalistic objection consists in the argument that biological phenomena are so complex and so difficult to control that teachings concerning their use and application lack the essential feature of repeatability and that they consequently do not have the required technical character to be regarded as inventions. Again, it seems that this assertion is rather outdated. First, there are more and more innovations in the field of biology which consist in or are based on a repeatable teaching and which are completely entitled to be protected as inventions. Secondly, the concept of repeatability is not such a rigid concept that genetic identity is required. The necessary degree of reproducibility depends on the nature and on the purposes of the invention.[17] "General reproducibility" may suffice to show technical character.

Another possible line of reasoning which strikes an emotional cord in some critics focuses on the indisputable fact that life is a phenomenon of nature which

---

15 Art. 52 (1) EPC, § 1 I PatG.

16 BGH GRUR 1969, 672 – "Rote Taube" = 1 IIC 136 (1970): "A historical interpretation of the term invention is even more unsatisfactory ... because invention is a basic concept in a field of law having as its most important task the encompassing of the patentable results of the most recent states of science and research. Therefore, the intent of the Patent Act itself not only permits but even compels drawing upon the latest state of scientific knowledge to interpret this concept of invention ...

17 Cf. TBA T 292/85, OJ EPO 1989, 275 = 20 IIC 725 – Polypeptide Expression/ GENENTECH I; TBA T 281/86, OJ EPO 202 = 20 IIC 726 (1989) – Preprothaumatin/ UNILEVER. Cf. also Szabo, Patent Protection of Biotechnological Inventions – European Perspectives, 21 IIC 468, 471 ff. (1990).

is not and cannot be invented by man. It is argued that to consider a biological innovation as an invention and to attribute it to an individual patent applicant by granting him an exclusive right, would ignore that this applicant is not the Supreme being and is only to a very small degree responsible for the existence of his invention and its insertion into nature. This view is, however, based on a general misconception of patent law: Every inventor has to use natural forces and preexisting components. It is equally obvious that every inventor stands on the shoulders of preceding inventors. Thus, for example, a new chemical process may be claimed and protected by a patent even if it includes only one additional step in a long chain of known steps.

Some critics accept that patents may be granted for the use and the application of biological processes, but argue that living organisms as such do not constitute patentable subject-matter. In order to support this assertion, reference is sometimes made to the exclusion of discoveries or to the so-called product-of-nature doctrine. These arguments, however, fail. Neither a microorganism whose genome has been changed by recombinant DNA-technology nor a newly bred plant variety nor a transgenic animal can be reasonably regarded as mere dicoveries. The real problem seems to be a terminological one. The language of patent law is, it is true, reminiscent of its historical origins when it was more or less restricted to inventions in physics or chemistry. The application of this language to biological inventions, especially to those consisting in genetically modified living organisms, may easily lead to the criticsm of reductionist or technocratic thinking which allegedly puts living organisms on the same footing as machines. Yet this criticism appears to be somewhat superficial or even unfair. If one accepts the proposition, which we think is correct and sound, that patent law has to be understood as the basic neutral instrument for the promotion of research and development, and if one accepts furthermore that man is allowed to use living organisms for his own purposes, as he has always done in history, this emotional barrier based on terminological superficialities should disappear.

Notwithstanding the fact that biological innovations may perfectly well fall under the concept of invention, it has been argued that the patent system was not designed to include biological inventions, so that their incorporation into patent law would lead to great difficulties. Problems are seen e.g. concerning the requirement of enabling disclosure, the determination of the scope of right and the exhaustion doctrine. To be clear, this position is not necessarily tantamount to denying protection to biological innovations altogether. On the contrary, the conclusion drawn may well be that there is a need for establishing a separate protection system allegedly better suited for biological inventions in order to complement or to replace patent law in this area of technology. This view has had the greatest impact in the legal treatment of plant inventions since it led to the creation of national plant breeders' rights systems and to the international system of UPOV, but it has also been advocated for other biological innovations.

In our opinion, one should not underestimate the capacity of the existing patent system to adapt its concepts to the challenges of new technologies. Certainly, there may be some types of biological innovations which face difficulties in being adequately protected under current patent law, and which nevertheless merit protection since they require substantial labour and investment. However, problems of this sort do not warrant the establishment of a completely new kind of industrial property protection. We already have a surfeit of such systems, e.g. for semiconductor chips or for type-faces.[18] The main threats consist in the practical and time-consuming difficulties in establishing such new systems, especially on the international level, in the tendency to over-regulate and in the splitting up of the international system of intellectual property as a whole, with the danger of atomization and the genertion of interface problems. So we would strongly argue against such a separate system for biological inventions. This does not mean, however, that one should abolish the plant breeders' right system, which has already a rather long tradition and is nowadays accepted by the interested circles. But it is important to view such a special protection scheme as a complement to the general patent law system rather than its counterpart. We will return to this issue in a moment when we discuss the exclusion of plant varieties from patent protection.

Our preliminary conclusion is: There is no explicit rule or concept in patent law that excludes innovations from protection for the sole reason of their biological nature. It would be neither sound nor realistic policy to attempt to change this state of the law. This is in line with Art. 2 (1) of the EC draft directive[19] according to which "a subject-matter of an invention shall not be considered unpatentable for the reason only that it is composed of, uses or is applied to biological material".

2.3. *Plant inventions*

European patent law contains an exclusion for plant varieties and for essentially biological processes for the production of plants (Art. 53 [b]). The same holds true for the harmonized patent laws of EPC member countries.[20] Fortunately, this exclusion is currently construed rather narrowly by the case law of the European Patent Office[21] and by the majority of legal commentators.[22]

According to our opinion, the underlying concept to reserve one area of possible inventions, namely plant varieties, for a special protection scheme, the plant

---

18 Fortunately, computer programs have found their legal basis into the existing copyright system.

19 COM (92) 589 fin. – SYN 159.

20 Under some national patent laws, the exclusion is restricted to those plant varieties which may be protected by plant breeders' right. This restriction was also provided for in German patent law until it was abolished in the context of the plant variety protection law reform of 1992.

21 TBA EPO, 1984 OJ EPO 112 – Propagating Material/CIBA GEIGY; TBA EPO, 1990 OJ EPO 71 = 21 IIC 361 (1990) – Hybrid Plants/LUBRIZOL.

22 For details cf. Moufang, Protection for Plant Breeding and Plant Varieties – A Frontier of Patent Law, 23 IIC 328, 343 ff. (1992); Neumeier (supra note 12), p. 181 ff; Straus, Biotechnologische Erfindungen – ihr Schutz und seine Grenzen, GRUR 1992, 252, 262.

breeders' right system is most questionable. It neglects the already mentioned basic principle that special protection systems should not restrict the scope of application of general patent law, but should only supplement it by additional protection for valuable innovations. In the United States, newly bred plants can be protected by general patent law and by two specific schemes which have been introduced by the Plant Patent Act of 1930 and by the Plant Variety Protection Act of 1970[23]. In the international context, the unfortunate double protection prohibition anchored in the old text of the UPOV convention has been abolished during the revision conference of 1991[24]. This has in principle opened the way for national legislators and for the EPC member states to delete the existing exclusonary rules in their patent systems. It would be welcomed if they could indeed take this step in the near future, although one has to acknowledge that the revision of a multilateral treaty such as the EPC is per se a very difficult task. The EC Draft Direktive takes the more realistic approach by accepting that we will have to continue to live, at least for a while, with the existing exclusionary rules, and limits itself to ensuring that they are interpreted in a restrictive manner. This is better than nothing, but it remains a preliminary second-best solution.

2.4. Animal inventions

The European Patent Convention – Art. 53 (b) – and the harmonized national patent laws also contain an exclusion for animal varieties and for essentially biological processes for the production of animals. Here again, the prevailing opinion in case law[25] and literature[26] favours a narrow construction of the provision. While accepting the very existence of the exclusion, the EC draft directive equally makes clear that it should not be read broadly[27].

There does not seem to be any justification for patent law to discriminate against inventors in the fields of animal breeding whether they work with more traditional methods or with genetic engineering techniques. Furthermore, the negative impact of an extensive interpretation of the exclusion would be even greater in this area than in the field of plant inventions because there is no supplementary protection system for animals. However, it is here that moral concerns have been raised by non-governmental organizations like Greenpeace and large segments of the general public. An indication of the level of opposition was given

---

23 Cf. Adler, Can Patents Coexist with Breeders' Rights?, Developments in U.S. and International Biotechnology Law, 17 IIC 195 ff. (1986) = Können Patente und Pflanzenzüchterrechte nebeneinander bestehen?, GRUR Int. 1988, 11 ff.

24 v. Pechmann/Straus (supra note 9), GRUR Int. 1991, 507, 508.

25 TBA EPO, 1990 OJ EPO 476 = 22 IIC 74 (1991) – Onco-Mouse/HARVARD.

26 For details cf. Moufang, Patentability of Genetic Inventions in Animals, 20 IIC 823, 832 ff. (1989); Straus, Ethische, rechtliche und wirtschaftliche Probleme des Patent- und Sortenschutzes für die biotechnologische Tierzüchtung und Tierproduktion, GRUR Int. 1990, 913, 920 ff.

27 Cf. Art. 3, 4, 6.

by the very recent emergency motion of the European Parliament tabled by the Green Group and approved by an overwhelming majority asking the EPO to revoke its Oncomouse patent.[28] It is our conviction that general ethical concerns against patents on animals are not well-founded. We love animals and do not like to see them suffering. But, as Baroness Warnock, the leading British moral philosopher recently stated: "I have never understood how patenting could do any harm to animals."[29] There are certainly some developments in modern animal breeding and related technology that need critical review. But the denial of patent protection is not the right answer, and patent offices as well as patent courts are not the most appropriate forums to handle these questions with ultimate safeguard provisions such as Art. 53 (c) EPC,[30] which excludes inventions the publication or exploitation of which would violate ordre public or morality. In its Oncomouse decision, the Technical Board of Appeal has required that the possible risks incurred by the exploitation of the invention be carefully weighed up against the benefits brought by the invention to humanity. The EC-draft directive intends to exclude processes for modifying the genetic identity of animals which are likely to inflict suffering or physical handicaps upon them without any benefit to man or animal. It is a common characteristic of both approaches to exclude animal breeding inventions for ethical reasons only under exceptional circumstances.

To sum up: It is unsound policy to deny animal inventions the status of patentable subject-matter. One should drop the exclusions of animal varieties and essentially biological processes for the production of animals altogether.

2.5. *Biological Inventions Concerning the Human Body and Its Parts*

More difficult problems arise with regard to the patentability of those inventions which are applied to or use the human body or its parts. It seems that, at least for some of these innovations, ethical questions of greater magnitude and complexity come into play and require a careful weighing up of the concerned interests.[31]

---

28 Cf. Thurston, Recent EC Developments in Biotechnology, (1993) EIPR 187, 188.

29 During the panel discussion Eposium 1992. Cf. Genetic Engineering – The New Challenge, Conference Proceedings and Essay Competition (ed. Cookson et al.), Munich 1993.

30 Cf. Memorandum of the Federal Ministry of Justice on the Revised EC-Draft Directive (May 1993), p. 12: "Ein wirkungsvoller Tierschutz kann nicht durch das Patentrecht gewährleistet werden, sondern setzt im Vorfeld bereits nach dem Tierschutzgesetz ein, wenn bereits in der Forschungs- und Entwicklungsphase, die der Erfindung vorausgeht, die Verwendung von Versuchstieren und die Durchführung von Tierversuchen nur eingeschränkt zulässig sind."

31 Cf. Appel, Der menschliche Körper im Patentrecht, in: Genetic Engineering – The New Challenge (supra note 29), S. 201 ff.; Dolder, Schranken der Patentierbarkeit biotechnologischer Erfindungen nach dem Europäischen Patentübereinkommen, Mitt. 1984, 1 ff.; Galloux, Ethique et brevet ou Le syndrome bioéthique, Dall. 1993, Chron. XIX, p. 83 ff.

The European Patent Convention expressly excludes methods of medical treatment by stating in Art. 52 (4) that they are not regarded as inventions which are susceptible of industrial application. This provision has a long tradition in most national patent laws, the reasons underlying it are of social-ethical nature. Whether these policy considerations are still convincing or not, remains open to debate.[32] Since medical and pharmaceutical products constitute patentable subject-matter under the EPC and the national law of its member states, the legislator has in principle recognized that patenting exemptions do little to improve the medical care of the public, and that, on the contrary, research and development have to be encouraged in order to improve the conditions of the public health sector. There are good reasons to replace the prohibition on patenting medical methods by a less drastic provision, e.g. by making the granting of a patent for a medical method dependent upon a declaration of agreement to license according to Sec. 23 German Patent Act (Lizenbereitschaftserklärung),[33] and to allow doctors to use a patented therapeutic or diagnostic method in case of urgency without obtaining prior permission.

However, with regard to the outcomes of human genome research, the exclusionary provision is, in its current form, not a serious patenting obstacle. According to the prevailing opinion, Art. 52 (4) EPC concerns only those medical method inventions that are carried out on a living body. The provision does not embrace diagnostic methods which are carried out upon tissue, organs or fluids which have been removed from the body permanently.[34] Furthermore, as it is explicitly clarified in the secon sentence of Art. 52 (4) EPC, products that are used in a medical method are not excluded.

While it may be worthwile to seriously consider the integration of methods of medical treatment into patent law, one should take a more critical stance against those developments which pose real threats to human dignity. In this respect, reference has been made primarily to the exclusion of inventions which violate ordre public or morality.[35] Furthermore, it has to be asked whether there is a need for additional safeguards attained by introducing provisions which are more precise. In general, ethical concerns may either focus on the invention claimed or on the patenting itself, i.e. on the granting of an exclusive industrial property right for a certain subject-matter. A few examples may serve as illustration:[36]

---

32 Cf. Beier, Future Problems of Patent Law, 3 IIC 423 (1972), at 441 ff; Moufang, Methods of Medical Treatment, 24 IIC 18 ff.; 47 ff. (1993).

33 Cf. Bernhardt/Kraßner, Lehrbuch des Patentrechts, 4th ed., Munich 1986, p. 130, note 69.

34 For details cf. Moufang, Methods of Medical Treatment Under Patent Law, 24 IIC 18, 36 (1993).

35 Cf. Beier/Straus, Gentechnologie und gewerblicher Rechtsschutz, in: 25 Jahre Bundespatentgericht, Festschrift (Hrsg.: Bundespatentgericht), Cologne etc. 1986, p. 133, 136.

36 For details cf. Moufang, Patentierung menschlicher Gene, Zellen und Körperteile?, GRUR Int. 1993, 439, 446 f.; Straus (supra note 22), GRUR 1992, 252, 257 f.

Certain applications of genetic engineering or reproductive biology to human beings have to be considered as contrary to the most fundamental values of our society. Some of them even constitute criminal offences under the legislation of several states. Cloning experiments, artificial twinning of embryos and the formation of chimeras are the most repugnant examples. Any attempt to obtain a patent on such innovations would have to fail since Art. 53 (a) EPC or the corresponding national provision would apply. The same holds true for inventions directed to gene therapy practised on human germ cells. Recently, the Examining Division of the EPO raised an objection based on ethical grounds against an application which concerned a genetic engineering invention dealing with the production of biologically active compounds in the milk of mammals.[37] Since some of the claims were directed to mammals in general without explicitly excluding human beings, Art. 53 (a) EPC was properly considered applicable. Inventions which genetically modify the human body in order to be able to use it as a "factory" of valuable substances will in most cases violate morality. They will also be caught by Art. 2 (3) (b) of the Revised EC Draft Directive according to which "processes" for modifying the genetic identity of the human body for a non-therapeutic purpose which is contrary to the dignity of man shall be unpatentable".

Whether the invention itself raises moral concerns or not, the granting of an exclusive right for the invention may well do. There can be no doubt that Art. 53 (a) EPC would block patents on human beings as such. Consequently, the abovementioned European patent application had partially to fail not only because the invention itself is repugnant when applied to humans, but also because the legal order can simply not accept property rights on human beings. Further examples of the scope of application of Art. 53 (a) EPC include patent claims on human embryos or fertilized egg cells or patent claims on human organs. In this respect, the concretization of this provision by Art. 2 (3) (a) of the Revised EC Draft Directive, which forbids the patenting of the human body or parts of the human body per se, is to be welcomed.

Most difficult problems arise, however, with respect to the patentability of human cell lines and human genes. Current patent practice has accepted exclusive rights on human somatic cell lines, hybridomas and genes, without invoking Art. 53 (a) EPC or corresponding national provisions. Should this practice be stopped? It is obvious that in general neither the research leading to such inventions nor their use is objectionable. On the contrary, the production of valuable substances like human interferon, tissue plasminogen activator or erythropoietin by genetic engineering with the help of human DNA-sequences is a highly desirable goal. Nevertheless, doubts as to the patentability of human body parts remain. Is it sound policy to encourage the increasing commercialization of human body parts or should patent law observe some caution? On the one hand, general civil law recognizes tangible property rights on body parts, e.g. organs, cells or fluids. On the other hand, the parallel between tangible and industrial property

---

37 European patent application No. 88301112 of 10 February 1988. Cf. GRUR 1992, 678 f.

law may be questioned since the grant of a patent for a certain subject-matter has a much clearer tendency to promote its commercialization than the recognition of a tangible property right. It is doubtful whether the forthcoming EC-directive intends to exclude human cells and genes from patentability. So far, the wording of Art. 2 (3) (a) of the Draft Directive is rather ambiguous and has already been the subject of several interpretations during the Travaux Préparatoires.[38] In our view, the legislator should determine more precisely the borderline between acceptable and unacceptable patent rights on human-related inventions.

3. Inventions Made at an Early Stage of the R & D Process

3.1. Part of the concern generated by the Craig Venter application within the scientific community and industry circles is due to the fact that the proteins coded by the claimed cDNA-sequences were not identified at the time the application was filed. It was asserted that taking out patents for such research results would lead to an exclusive mining licence in an unexplored field. It seems to us that this argument has certain merits, although it should be presented in a more balanced manner.

3.2. In the last decades, patent law has in fact developed in such a way that protection can be obtained by inventors acting in relatively early stages of the technological R & D process. The conceptual distinction between invention and discovery is no longer of great importance, since it is well established that a discovery may be the very core of a patentable invention and may also be considered when assessing the requirement of inventive activity. Consequently, genes that have been isolated from nature can generally be patented although, in such a case, the scope of protection does not include the use of the gene in its natural environment. Product protection for chemical substances is granted in an absolute manner, with the result that the patentee can exclude others from performing even those uses of the product that he himself has neither foreseen nor disclosed. The criterion of industrial applicability is defined in a very broad manner in Art. 57 EPC and can be met by almost all patent applications. Thus even products and processes which can only be applied in research may perfectly well be the subject-matter of a valid patent.

3.3. Critics form industry circles view the NIH applications an cDNAs as an extreme example of a widespread practice in biotechnology that seeks to control not inventions, but the means of making inventions.[39] One commentator observed:

---

38 Cf. the Explanatory Memorandum of the proposed Directive, p. 7; contra: the comments made by the Rapporteur Rothley during the debates in the European Parliament (Report of the Parliamentary Session of 26 October 1992, p. 69).

39 Kiley (supra note 1), 257 Science 915 (1992).

"Patents are being sought daily on insubstantial advances far removed from the marketplace. These patents cluster around the earliest imaginable observations on the long road toward practical benefit, while seeking to control what lies at the end of it."[40]

3.4. In our opinion, it would be sound policy that the patent law encompasses research results which are "close to industrial application" (anwendungsnahe Forschungsergebnisse").[41] Otherwise, patent protection would come into play at too late a point in the innovation process, namely, at the point when the fruits of applied research have entered the development stage. At the Max-Planck-Institute, we have proposed nearly twenty years ago that one should no loger require that the invention be commercially applicable at the filing date of the application, but that one should open the doors to patent protection also for such concrete research results which can, in all probability, be expected to be developed into industrially exploitable inventions in the foreseeable future.[42] At the same time, however, we warned against granting protection at too early a stage, namely for pure knowledge gained during the basic research phase, since it would result in protection which would be too broad, thus having wide-ranging deterrent effects.[43] In fact, patent law is not an entirely appropriate instrument for the promotion of necessary basic research.

3.5. Notwithstanding our general sympathy for the protection of science-based innovations, one has to recognize that current patent law sometimes shows a tendency of over-protection of early inventions. This may have a stifling effect. When a patent applicant is allowed to cover a broad spectrum of indefinite possible uses of his research results that have no real basis in the technical teaching disclosed, it results in the discouragement of competitors who may also have invested heavily in research and development.

Complex innovative processes have to be stimulated by patent law in all relevant stages. Although pioneering inventions that open up completly new fields of technology merit strong protection,[44] the basic rule has to be respected: No one has the right to obtain a patent for mere speculation. In order to prevent such exclusive rights from coming into existence, the disclosure requirement has to be applied in a straightforward way. Relaxing this requirement might have a counter-productive effect placing competitors which sought to present a correct and fully enabling disclosure at a disavantage. For the same reason, broad generalizing claims have to be scrutinized very carefully, since due to the complexity

---

40 Ibidem.

41 For details cf. Beier/Straus, Der Schutz wissenschaftlicher Forschungsergebnisse, Weinheim/etc. 1982, p. 66 ff.

42 Cf. Beier, Scientific Research, Patent Protection and Innovation, 6 IIC 367, 378 f. (1975).

43 Beier, 6 IIC 367, 378 (1976).

44 For details cf. Welte, Der Schutz von Pioniererfindungen, Cologne/etc. 1991.

of biological phenomena, such claims frequently cover subject-matter for which the description as a whole does not provide a sufficient basis.

3.6. Patents granted on early results of the innovative process might block the exploitation of forthcoming developments. In general, industrial property lawyers rely on market forces to overcome situations in which two or more patents block each other. For instance, the production of complex machines "of the state of the art" would rarely be possible without a web of licensing agreements. Most patent systems contain provisions which allow the granting of compulsory licences in the public interest. However, such licences must be considered as an exception and will only be available under strict conditions.[45] In order to give a further incentive to deblock the exploitation of inventions made at a later stage in the R & D process, the legislation of most EPC member states facilitates the obtaining of a compulsory licence when the person applying for it is the owner or licensee of a dependent invention.[46] It seems worthwhile to think about the introduction of such dependency licensing rules into German patent law, provided that the dependent invention constitutes a significant improvement over the dominant invention or contains at least a technically important new step.[47]

4. Patent Law and Academic Research

4.1. Members of the scientific community including many scientists attending this workshop have expressed their concern regarding the detrimental effects of patenting on free communication and scientific exchange. For Academia, there is a fear that the gold rush mentality may develop. Yet the true picture appears to be different. To summarize our position, inventions should be protected in such a way that not only the results of industry research, but also the results of publicly funded research will find their way to the market of innovation and lead to technological and social progress.

Fifteen years ago, the German Ministry of Research and Technology commissioned the Max-Planck-Institute to carry out a comparative study on the relationship between the patent system and the governmental promotion of R & D. The findings of this study which included the USA, the Unitec Kingdom, France, Holland, Sweden and Germany were published in 1982 in three volumimous

---

[45] Cf. Beier (supra note 3), GRUR 1992, 228, 234. In the last decades, no compulsory licences were granted in Germany, with the notable exception of a recent decision by the Patent Court, Mitt. 1991, 243 – "Human-Immuninterferon".

[46] For details cf. Straus, Zur Zulässigkeit klinischer Untersuchungen am Gegenstand abhängiger Verbesserungserfindungen, GRUR 1993, 308, 312 f.

[47] Cf. Intellectual Property in Genome Mapping Programmes – A Report of an EC Workshop, 20th–22nd November, 1992, p. 16.

books.[48] They clearly indicated that universities and other research institutions should adopt a more active patent and license policy than before.[49]

There are still many scientists as well as government officials, who assert that taking out patents for publicly funded research is not necessary and even inappropriate because there is no need for publicly financed research institutions to get R & D expenditures reimbursed with the aid of patent protection, and secondly, inventions financed by the general public through tax money must not be monopolized but must be freely available to everyone. These often-heard arguments are, however, in conflict with basic patent theory as well as which practical experience. They overlook first that the exclusive patent right functions not only in the R & D phase of the innovation process. It also significantly advances the industrial application and dissemination of new technical knowledge, which is decisive for achieving technical and economic progress. It is only by securing the exclusive use of the invention to the patentee that the invention can be assigned and licensed to others without risk. The patent thus becomes one of the most important instruments for transferring technology; and here, in that transferring function, lies the main importance of the patent system for publicly funded inventions.

Experience teaches that an effective transfer of technology from the research laboratory to industry can be obtained only through an active and well oragnized patent and licensing policy. Even though it would appear to be appropriate to make available R & D results financed by government to everyone, by means of scientific publications, special information systems and other methods, these approaches have not been successful in practice. Nobody is really interested in technology which is freely available to everybody. The transfer of government-financed R & D results to industry has functioned satisfactorily only when the government or the research center has acquired a patent which can be exploited, as in industry, by licensing agreements. In fact, exclusive licenses even had to be offered, as a rule, in order to find suitable licensees willing to assume the necessary development costs and the risk of introducing the new product on the market.

4.2. In order to promote a fruitful coexistence of patent law and academic research, it is necessary, however, to ensure that the research process itself is not unduly restricted by patent rights. Most national patent laws in Europa contain an explicit research or experimental use exemption. The exact metes and bounds of this exemption are not wholly clear, as demonstrated by the case law in the very much disputed area of clinical trials. But at the risk of oversimplifying, it may be said that research on patented subject-matter is free, while (non-private) research with the help of the patented subject-matter falls under the prerogatives of the patentee.[50] Scientific research is therefore not completely exempt from patent

---

48 Beier/Ullrich (ed.), Staatliche Forschungsförderung und Patentschutz, 3 vol., Weinheim 1982.

49 Cf. Beier, Governmental Promotion of Innovation and the Patent System, 13 IIC 545, 561 ff. (1982).

50 For details cf. Chrocziel, Die Benutzung patentierter Erfindungen zu Versuchs- und Forschungszwecken, Cologne etc. 1986.

rights. As the example of PCR-technology illustrates, the danger exists that scientists may be restricted from using important research tools. There is a good case for clarifying and – possibly – broadening the research exemptions, perhaps subject to a financial remuneration for the patentee.

5. Conclusion

5.1. In biotechnology as well as in other areas of technology, an adequate industrial property regime bears great importance for realising the opportunities innovations may hold for society. Patent law is certainly not the only legal instrument for fostering technical progress. It is also not an appropriate instrument to control and prevent the socially undesirable consequences of modern technologies such as genetic engineering. One of its great advantages is its neutrality, in the sense that the patent system does not favour specific inventions and does not discriminate against others. It leaves the control over desirable or undesirable social consequences of inventions to the general legislator and the ultimate decision on the economic value of an invention to the market.

5.2. To sum up: An international understanding concerning the patentability of biological organisms and human genes should be developed along the following lines:

First, patent exemptions for specific fields of technical innovation constitute the ultima ratio of patent law and compelling reasons are needed to justify them. There are no such reasons with respect to biological inventions including product inventions, whether microorganisms, plants or animals. The human body or its parts as such should never become the object of patent protection. Caution is necessary with respect to inventions that use the human body or its parts for producing valuable substances.

Second, the protection of inventions or discoveries which are located at a rather early stage of the innovation process of research and development (= research results which are "close to industrial application") has to be welcomed in principle. There is, however, a current tendency of over-protection that may have a stifling effect. Possible remedies include a stricter application of the disclosure requirement and the legislative introduction of dependency licenses.

Third, inventions should be protected in such a way that not only the results of industrial research, but also the results of publicly funded research will find their way to the market and thus lead to economic and social progress. However, the research process itself should not be unduly restricted by patent rights. There is a good case for clarifying and – possibly – broadening the research exemptions, which are already provided in most national patent laws, in order to promote a fruitful coexistence of patent law and academic research.

What Does Pharmaceutical Industry Expect of Patenting Human Genes and Living Organisms?

B. Garthoff

In a world of fears regarding gene technology where a novel, respectively film such as "Jurassic Parc" (by M. Crichton, [1]) determine peoples' attitude towards gene sequencing, the rational discours of whether to patent human genes or those of living organisms seems almost impossible.

However, industry such as the pharmaceutical business in which patents play a dominant role, has to consider both aspects, i.e. the patentability of its efforts *and* the acceptance in the population. In that regard an adequate patent practice and rational patent jurisdiction not governed by emotions in basis and cornerstone of industries security in its business. What are industries views and expectations and which are the Dooms Day scenarios for the patent oriented corporations in patenting human genes or living organisms?

In general, patenting from the industry point of view is clearcut a blessing for both, the scientific and economic progress. It does protect intellectual property rights and therefore, guarantees that investments into science and technology pay off. And on the other hand, it does not prevent useful information and knowhow being used and disseminated – as it is examplified by the research use of recombinant DNA (Cohen Boyer patent [2]) and PCR technologies [3] – moreover, it ensures usage. In that regard, fears of some scientists raised during this meeting are not substantial – it cannot be in industries interest to prevent research scientists from using recent technology, rather induce them to cooperate and pursue joint research projects which might lead in the end to further scientific progress and additional benefit, eventually joint improvement patents. In this regard, patenting of human genes and living organisms is in no way different from patenting chemically synthetized substances and way of procedures. Similarly, strictly scientifically speaking, gene sequences are "natural substances with a chemical code" that are patentable in that one can describe the innovative way of finding and "producing" them. By this strictly rational approach, there is also no difference between human, animal or other living organism pattern of genes – and from an industry point of view, there cannot be a difference whether or not to patent either. In this regard, one also has to be aware that it can be expected that the difference in genetics between mankind and e.g. chimpanzee may be only about 1.6 %, all the rest being identical (according to the pysiologist Jared Diamond [4]).

Even with the Drosophila fly we still might have 50 % genetic code in common. So even if one would like to differentiate, in the process of patenting, we might end up with the same sequence of e.g. Factor VIII gene in several species including man and dog. Therefore, how would you differentiate in a scientific world just for the patent process human and diverse species if the code is exactly the same? From a science point of view, one has to admit, you cannot. And as current patent practice teaches, patenting of genes such as the Factor VIII genes [5] or transgenic animals has been widely accepted, although not unequivocally [6]. The apparently next question that arises then; can the whole man by virtue of patenting the human genome be patented? An interesting article by Baumbach and Rasch [7] analysed this aspect and concludes – although this might theoretically be construed – that this is not to happen for different reasons and in the best "or worst" case, there could only be 1 % of the whole genome at a given time be patented.

What are the gloomy scenarios for pharmaceutical industry?

Number 1:

That patents be allowed in some countries for cDNA sequences with no obvious function at time of filing.

To be subject of a patent, an invention has to be new, must have an utility, and must improve that state of the art. This should also be true for patents of human genes and organisms. Thus it is questionable whether the thousands of cDNA sequencies (2700), that the NIH wanted to file patent applications for, really are patentable. The U.S. Patent and Trademark Office rejected the NIH application last year. The utility of DNA sequences, whose function is unknown, is not obvious, even if they may comprise gene fragments. From a pharmaceutical industry point of view this application could not be allowed, and any similar in future has to be rejected. Approval would contradict also patent practice of the past in chemistry and biology. On the other hand, the sequence of a whole gene, whose gene product is known and may have a useful application – whether as a therapeutic agent, or as an ingredient of a diagnostic kit, or as a tool in an experimental setup –, fulfills the above cited criteria, and patentability should be possible. The argument, that genes already existed in nature for thousands of millions of years, and thus could not be invented, does not hold, since the deciffering of the gene sequence and its useful application, improving the state of the art, is the intellectual achievement that leads to the patent. A native gene exists of many more than the DNA bases that are necessary for the mere coding of a protein. There are introns which are not transscibed, there are TATA boxes, promotors, enhancers, and other regulatory elements. To have a useful gene application, intellectual work is necessary to put the right sequences together, excise introns, select appropriate promotors, and so forth.

Just as an example, the Bayer US subsidiary Miles produced for a long time the blood clotting factor VIII for hemophilics from plasma of blood donors. It now owns the right for the human factor VIII gene and produces and markets human recombinant factor VIII already in the States. Producing human recombinant proteins for therapeutic purposes in microorganism and cell culture is without any doubt a significant improvement of the state of the art, and an impressive achievement in safety, since one is able to produce sufficiently large amounts of life saving substances without the risk of contamination.

Coming back to what is patentable; there is a controversy between Genetics Institute that kept broad claims to a gene fragment, and Genentech that later claimed the whole gene sequence of factor VIII (and from which Miles received the rights). In the interference proceeding between Genetics Institute and Genentech, Genentech is the senior party because they were first to disclose the full length sequence. It is expected that the U.S. Patent Office rules in favor of Genentech in the respective interference proceeding (n° 102, 331). The significance of such a ruling would be that the U.S. Patent Office concluded that there was simultaneously conception and reduction to practice at the time the full sequence was known – and this is in keeping with the earlier ruling in the Amgen case regarding erythropoietin. It is clear that a gene fragment does not in and of itself enable one skilled in the art to determine the full length gene. Even substantial experimentation does not guarantee that one arrives at the full length gene or a functional gene.

In consequence, you can now also use the patented factor VIII gene in an approach for gene therapy in hemophilic patients that lack that gene.

Number 2:

That moral and economical (= i.e. patent) aspect are mixed.

Whether to patent fragment cDNA sequences out of nature (no matter animal or human) is a decision of society and politics and underlies some kind of political risk-benefit evaluation. We cannot dump the decision on patent officers of e.g. the EPO (European Patent Office), as this is not the appropriate instance. I strongly object to the view that – as long as there is no political decision – EPO should substitute and rule in advance and instead. That is stretching Article b) 53 EPU in EC regulations too far.

Number 3:

That there might be varying ruling or different laws in the future with regard to patenting sequences or with regard to such “things” like "onco-mice”.

We do have patented sequences worldwide, already. If there were changes in laws, one would have to make sure that these changes are harmonized and orchestrated worldwide in the same way. Otherwise we would find ourselves in the same situation as before – industry and science in Europe, and specifically Germany, lagging behind in technologies due to restrictive law making. One has to remember that although patenting is European, exploitation of patents come under local law of the member states.

UK research institutions have followed the NIH step and startet filing for patenting of sequences. Industry is concerned that we have differences worldwide which put disadvantages to our expert oriented industry and therefore, to our society that lives on it. As mentioned before: predictability for exploitation of investments into patentable matter is the business basis of this industry and a guarantee for worldwide harmonised laws in this respect would be a must.

And finally, there are some other areas in this field that will be controversial. What about e.g. when one country in the EC accepts patents interfering with the germ line what most likely will be totally excluded in others?

And how about the use of transgenic animals for production of urgently needed remedies such as endogenous factors (e.g. $_{\alpha 1}$PI) where local laws such as animal protection legislation might be controversial in different countries and prevent this usage?

It has to be agreed upon that there must be given the chance to industry to invent and patent together with sciencists at university for the societies benefit by investing and putting such patents into effect worldwide (and only those things that are useful will be paid for by community anyhow). The willingness in industry is there to cooperate.

We already do have disadvantages vs. some countries, e.g. USA in so far, as an European company might have to reduplicate its findings on US territory in order to get priority on its patents in general.

Summary

The difficulty for pharmaceutical industry with regard to the somewhat unpredictable situation in patenting human cDNA sequences or those of living organisms is characterized. Worst cause scenarios for the future development in this area are given and the significance of predictability of economic exploitation of investments in such patents is emphasized. Cooperation with academic and scientific research at universities is called for.

References

1. Crichton M (1991) Jurassic Parc. Ballantine Books, New York
2. Cohen SN, Boyer HW (1990) US Patents 4237224, 4468464 (1984), and 7470470 (1988)
3. Mullis et al. (1990) (Cetus Corp.) Process for amplyfying, detecting and/or cloning nucleic acid sequences. US patent 4 683 195
4. Diamond J (1993) (UCLA) The Third Chimpanzee "The Great Ape Project", pp 88–101. Editey by Cavalieri P, Singer P. Fourth Estate, London
5. Capon DJ, Vebar GA, Lawn RM, Wood WI (1985) (Genentech Inc.) New recombinant functional human factor VIII or derivatives – useful for treating haemophilia, and obtained as pure product by recombinant DNA technology EP patent 160 457
6. Kinkeldey U (1993) Die Patentierung von Tieren. GRUR Int Heft 5:394–399
7. Baumbach F, Rasch D (1992) Kann man das menschliche Genom und damit den Menschen patentieren? Mitteilungen der deutschen Patentanwälte 83:209–212

Acknowledgement: Assistance in preparing the manuscript is gratefully acknowledged; Dr. G. Thomas, Bayer AG
P. Simonton, Miles Inc., Pharmaceutical Division

Discussion: Session 6

Hofschneider: I would suggest that we organize the discussion in the following way: We first could have a discussion between the three speakers, and then we could have a discussion with the audience, including points of view of medicine, science and industry.

Beyreuther: I think it's a valid point. What comes out to my surprise here is that industry more or less follows the rules that we have accepted in science – very little legislation and publish as fast a possible and as precisely as possible. This is an attitude that has changed and it's not an attitude that's been taken up by German industry completely. Maybe the Swiss – I know that the younger generation is doing it. I can quote a very interesting remark of one of the colleagues from nearby who said, if I give you the information on this I only help my competitor. So, he didn't give me the strain. It was very recently. If this is becoming the general practice, and I know that in the States it's also becoming the general practice, we can more or less agree on this – as long as we publish fast, as long as we keep legislation only as stringent as necessary, that's the best thing for science and industry. And as long as we have the very open dialogue we had here, I agree with you fully. I still think that we haven't paid attention to the evolutionary implications of gene patenting and the implications for human genetics but there is no precedent that could be translated into misuse or so. I would be happy if the acedemic need for more clarity in regard to freedom for use of patented material, if this could be stated more formally by the industry, that would help very much to take away all our problems. For instance, the PCR technology – nobody is going to clone a gene without using PCR. I agree and also with Dr. Ganten: I have no problems that there are maybe a handfull of animal models that may be used, for example, as donors for organs, or a handfull of cells that will be implanted into the brain. They need to be patented because they have to have certain receptors in order to develop in brain into neurons. There may also be a handfull of species – cells and animals – which need to be patented but I agree with Dr. Rowland: In order to get a patent, you have to define the animal by defining the integration site. You have to say in which chromosome the gene is integrated. If anybody has a different strain, it won't fall under this patent. You are never able to define an animal on the basic of its genome, you have allelic differences. If you say this animal is defined by the integration of this gene at a certain site, I have no problems in accepting a patent.

Ganten: I don't see any major disagreement with what I said and what has been said from the point of view of pharmaceutical companies. I think it is a general problem of how our societies organize themselves. We try to legalize procedures and foresee by laws what might develop in the future. I think it would be much better to have a minimum of legalization and of regulatory rules and then react in a very pragmatic way. I think it's an important point especially in the gene technology area, – something which probably has not been done sufficiently in Germany but in Switzerland and I think also in the States and certainly in France: People actively involved in the biotechnology field go public. They talk about what's going on, don't do it only in the labs. I think they keep society informed, and don't create the probably wrong impression that there is something mysterious going on in the laboratories which usually, one can say, unfortunately is not. Usually laboratory experiments care much more down to earth than most people think, apply basic, well tested methodology but the real inventions are rare. To try to reach a consensus between the scientific community and society at large, public information the main thing I regard as necessary.

Hofschneider: Yesterday we agreed that one should not have a risk benefit discussion in the context of patenting. According to what you said today, I'm not so sure if you want to have such a discussion in the context of patenting or not. Then there would be a disagreement with some people here.

Ganten: It depends on what you call a risk benefit. I was alluding to new treatment modalities by any kind of method, whether gene technology, of course, or others. If you want to introduce a new drug, patented or not, you have to consider the risk of introducing this drug with all the unknown consequences – that's what I meant.

Yorke: I really don't think I have any more to add. I've certainly noted very carefully and meant what I said about what is an infringement really as seen from the researcher's point of view. As regards the speed of this process in human gene therapy, I would have to think about that, I haven't got a ready answer to that problem.

Hofschneider: I personally would like to add two other points which I don't find covered by your list (addressing Yorke) and that is the demand of Dr. Beyreuther to have a copyright in addition to a patent. Maybe it would be useful to discuss here what such a copyright should mean. Then we should discuss a grace period. You made a statement which was a little bit surprising for me. You said that to get an invention patented in the United States you have to make that invention first in the United States. Did I understand you correctly? And if that is so, then we should discuss whether that guarantees fairplay.

Yorke: No, it's not quite so. Outside the United States the person who wins a patent conflict is the person who filed a proper patent application first. So, if we file two weeks before Bayer on the same compound, then we win. In the United States this isn't true. It has to do with who invented it, you have to go back to the laboratory notebooks, conception of the invention, all sorts of things. On each of those you can argue for years. The grace period point comes up in the harmonization discussions with America where the rest of the world would obviously like the United States to come to first-to-file but America would like us to introduce a grace period.

Beier: Just to the grace period. I think it does not come up only in the struggle with the Americans, it came up much earlier, namely at the time at which the German legislation unfortunately gave up the grace period of six months. Since then we have been fighting for the reintroduction of the grace period especially in the interest of the scientist. And it is understandable, Dr. Yorke, that a head of an industrial patent department says that he doesn't need a grace period because he has everything under control, he has his inventors under control, they would never publish before the patent department has seen the paper but in reality I know that it's not the case, especially in our field here where industry cooperates with universities and university inventors who don't have the discipline as the inventors employed by large companies are deemed to have. So, I think first it was understandable that the industry, – large industry, because the small and medium industry is never represented, – was against the grace period because they thought they don't need it. But there was a change in opinion some years ago and Dr. Hüni, your colleague from Ciba-Geigy, was the first to publish that change of opinion in our journal. That was the question of the grace period. Now it is indeed a bargaining tool with the Americans and I hope that even those countries, Sweden, France, which have been against such a grace period for theoretical reasons, would also adopt this. If you permit me, Mr. Chairman, to say some words about your thinking that today the patent specification would not contain a sufficient disclosure.

Hofschneider: Would you allow me to finish one other point first. Dr. Beyreuther, do you agree with the idea that if you have a grace period you would not need a special copyright regulation?

Beyreuther: I think 20 years ago I thought a little about the requirement of sufficient disclosure and I came to the conclusion that today, other than in the 19th century, you could describe an invention on one or two pages and that was a sufficient disclosure which enabled the third parties to work with the invention. But today even with 30, 40, 50, pages of patent specification, you are not able to produce the product or to construct the machine. You need additional information from the inventor. Therefore, you have to conclude a special agreement to get the related know-how in the field of life science; maybe an additional agreement to get the probes because they could not sufficiently be described. There is in my opinion a change in the function of disclosure which we have to accept. On the other hand, the access to technical information through data banks is ameliorated.

The general information on the state of the art is good – in connection with the fact that most countries publish patent specification at least 18 months after application. But it is very, very, rare that you can take a patent specification and give it to your engineer and say, O.K., in two months we will imitate the invention. As long as the patent is protected you cannot imitate but the information which is given in this patent specification is sufficient for looking for alternative technologies, for circumventing the invention.

Beier: As you have heard, it was not only my point, but for example, also Dr. Beyreuther's point and since I guide the discussion I should be neutral and so would you respond.

Beyreuther: I'm in a very competitive field and I see a lot of these patent applications. What I realize is that they have one page on the invention and 30 pages on what they want to do with it. This is usually a lot of hot air. These are not European patents – there's almost no European activity in the field I'm in. If you compare the background that is given on the invention with the general state of the art, the method section allows you to evaluate whether the statements being made are correct or not. In a patent application you state what you have and you usually do not state the way you came up with the information. It's always said that the background paper can be ordered from the inventor. It happened to me that I got a sequence from the inventor which was wrong; it was my sequence, and I knew that it was wrong. I don't understand how this could be accepted by the patent office. It may happen very often that the sequence is wrong and it may be interesting in the future whether these patents can be upheld or not.

Straus: I first have one question and then one remark. The question first: Is it different with the peer reviewed papers? Can you reproduce all of them? Now my remark: the repeatability of descriptions in patent applications is no doubt a problem. I was recently inudued in a case in which this issue is under dispute for about 15 years. Many institutes were involved in testing experiments despite the fact that it's a quite simple invention. It's a microporous membrane used for cold sterilization, widely used now. But it's up to the patent offices also to tackle it in a more rigorous way.

Hess: To this point I have been involved and I was asked by an editorial committee from a very good scientific journal to repeat experiments prior to publication from a paper which was submitted. This happens all over again. So the peer review system is extremely rigid in good journals. It took me about two weeks to repeat the experiments. I think good review systems follow this pattern. This was actually an American editorial committee which asked for this and I think it's a very interesting example how rigid peer review system can work.

Yorke: If I could perhaps make a comment. I'd look pretty silly, I would have egg all over my face if after Sandoz had invested 200 or 300 million dollars, we filed a patent, we tried to enforce it, and the imitator shows that our patent application wasn't sufficient. One of the primary things all the people in our department have to do to is to try and make sure that that attack can't be brought against us. So it shouldn't happen. It can happen for various reasons, sometimes with smaller companies; it depends a bit on their patent attorney.

Garthoff: Maybe I'll also comment on whether a good quality paper is never revoked. Actually, it has happened even after publication in the area of Alzheimer's disease. There is peer review of all the competitor companies to some extent. Of course, if Sandoz comes up with a patent and everybody says that's totally wrong you're immediately going to fight against that. So, I think the peer review takes place. One additional remark: I feel a little bit beaten up in terms of German industry, specifically with those two groups with which we have been interacting. I just can assure you it's not the attitude just of the younger scientific personnel in German industry to widely cooperate with the science community in Germany and other countries as well.

Rowland: I would just like to say I think you're misconstruing the function of a patent application or a patent. We are supposed to tell you how to practice the invention not how we got there. My attitude is that for a patent application on a sequence all I should have to publish is the sequence and not tell you anything about how I isolated that DNA, what kind of probe I used. In the United States as distinguished from Europe the examiner can ask for all that information and demand proof that won't be in the patent, so that if you want to find out more about the invention you have to order the file history after the patent is issued. The other thing is exchange between corporations and academia. You have to realize that corporations are trying to make a profit and they just cannot give you their biologicals so you can pass them on to your friends. It's unfortunate, but biologicals would not otherwise exist but for the fact that there is a corporation and it did spend the money to make them.

Grunwald: From a practical point of view I would try to discuss the offer Mr. Yorke made. He made the offer to clarify the interaction between science and industry and I think it's a fair offer and I would like to know how we best could go about it. The second point, too, from a very practical point of view: We seem to be in accordance and all agree that a grace period would be of great importance for science and research and I would like to actively push the point. We did so a couple of years ago without success. Another point: I think there was quite an agreement on the points we discussed. We discussed and all accepted that tools should be patented, and as far as the procedure in deciphering genes in concerned there's no debate about that. I think the rest of the debate is still going on as to the finding itself, to the substance found. One last point, I agree that legislation shouldn't be invoked too early. On the other hand my fear is that the question of patenting mammalians will be a case of litigation for the next couple of years and this would in my mind develop the question whether we shouldn't move in some way, perhaps by a declaration of scientists, to discuss the principles under which an industrial property right in mammalians, especially as tools for human diseases should be protected.

Vogel: I wish to reconfirm two statements of Dr. Beyreuther. The first statement he made is that if genes are patented at all, then the limits of the contents of the patent should be kept very narrow and I think this coincides with something I said yesterday, that it should be possible to give patents for things like this just for a special purpose; maybe a gene sequence to make a certain protein and not for everything else. The second point you have made which I also wish to reinforce is that there the very well known conditions of novelty and non-obviousness of something that has to be patented. Now the technology in this field has been developed so far that identification, isolation of genes a.s.o. may be laborious but follows a certain fixed pattern. This means that the non-obviousness is not really given in most cases and that one should ask for really novel and unexpected procedures or novel and unexpected ways of coming to a certain solution. I think these are two conditions which I would regard as very important. This is my question to these specialists in patenting law and maybe Dr. Beier will answer this question in his contribution – whether these two conditions actually need new legislature which would be very bad because we live in a society in which life is covered by a net of mostly unnecessary legal regulations. Can these two conditions be fulfilled by suitable interpretation and, in my opininon, obvious and straightforward interpretation of the existing law? I only wish to make one remark to Dr. Yorke. What you said regarding developing countries, I think might certainly be true concerning pharmaceutical and biochemical substances. I have the feeling that the situation in seed production or even animal production might be quite different. One should consider this very carefully before any action is taken, for example, pressure against the intention of the developing countries themselves because we always have to assume that they know their own problems best and that we cannot intend to know their problems better than they do. The important point that two things Dr. Beyreuther has said, can these things be accomodated with in existing laws?

Sahai: I address myself to Dr. Yorke. There are three things. You mentioned that you were unaware of what the Indian farmers had been doing for the last 200 years. I'd like to tell you of one. The Indian farmers traditionally are seed producers. In fact, to give you very specific statistics, over 60 % of the seed requirements of Indian agriculture are met by Indian farmers through inter-farmer sales. That is decidedly one very traditional and very old activity that would cease to exist if intellectual properties in the form of patents were exercised there. Your second contention that patents do no harm to developing countries – I know of no study that has studied this, I imagine that it lies in the realm of conjecture. As against that I would like you to recall one slide which I had about a World Bank study that had assessed the effect of plant breeders' rights in agriculture in the United States. It showed a very dramatic shift from public to private sectors, free flow of information not in both directions, but only from the university to the commercial sector, but not from the commercial sector to the university or the public domain. These will have consequences for a country where all of this lies in the public domain. These would definitely have consequences and therefore, your conjecture that it would not harm developing countries – I'm not so sure. Third, your comment about patents in the pharmaceutical sector. I think you are aware that the Indian Patent Act allows for process patents, not product patents. There's the provision for process patents which has allowed alternative methods of production. I am possibly blue in the face from repeating but I will do it again – that product patents have been relatively new introductions in the chemical and pharmaceutical sector even in Europe. So, in India, it is something that is protected through the process route and not through the product route as has been the case in several countries in Europe as well.

Hofschneider: May I ask both parties not to become too specialized. I think everybody here has realized that there is a big problem, but maybe to solve it you need really specialized knowledge which we don't have available. Just now I see there is a problem that a world-wide rule should be changed to please the Indian situation or the Indian situation has been changed to fit into the world-wide regulation. I don't know which is the correct answer, so please make a short statement.

Yorke: First, the point about pharmaceutical patents is not whether they're products or processes. In India, their term is so short that there is no effective protection. It's just an imitative industry with processes; they will do no real research, there's no incentive. Secondly, Italy, you should study what's happened to Italy in the 10 years since they introduced patent protection for pharmaceuticals. You would learn a lot from that. How that has improved Italy in many ways – the pharmaceutical industry and the health sector. All the studies that you refer to are mostly from the 1960ies, written from a socialist point ov view. I have spent alot of time in the developing world and I would like to see the proof of any damage. It's easy to allege, but I really would like to see the proof.

Folman: We didn't really have enough time to comment on Dr. Sahai's important lecture. The point is: Different countries of the world amit are in different stages of development having to deal with problems which are absolutely different. Developing countries cannot fight the patent laws of the developed countries. What they should really do – and can do – is try to use as much as possible of the new knowledge to their advantage. They can use the public domain sector which has to remain the most important sector in research in their countries to the benefit of their farmers by trying to purchase the new seeds and the new technologies from the western world and thus improving the lot of the local farmers. Patents are for a limited time; after a certain period, they'll acquire the new technologies.

Gugerell: I would like to comment on a question from Prof. Vogel. Prof. Vogel has asked whether it would be necessary to implement new legislation on novelty and inventivity and my opinion, without anticipating any opinion Prof. Beier may want to express later, is that this is clearly not the case. Present law is able to adapt to development and I think what Prof. Vogel fears is that the assessment of the inventive step is not properly done but the remedy for that is: The assessment of the inventive step is done in patent offices and by judges; therefore, the remedy can only be to train the people who do the job so that a meaningful result is achieved. At the European Patent Office I think we have achieved a good standard.

Round Table Discussion

Hess: I would first like to ask Prof. Zimmerli to answer a number of questions which might come from the audience with respect to the ethical problem.

Zimmerli: I feel even more convinced than I did this morning that the important ethical issues which still exist are not related to the questions of patentability because all the important ethical questions whether related to our responsibility to the Third World, whether related to the question of responsibility to future generations, whether related to the question of our responsibility to non-human nature or related to the question of our responsibility to our fellow beings, – are questions which still exist even if we do answer the question whether we should patent living organisms and even parts of human beings positively. So, the ethical questions still exist, but the question of patentability is not specific to the ethical realm. My conclusion is that we have to look at each argument very carefully and to distinguish between the ethically relevant questions in order not to lose our battles on the wrong battle fields. But we should rather concentrate on questions like responsibility to Third World, responsibility as opposed to future generations, discuss these questions and not invest too much of our energy and our intellect in the question of whether or not we should patent genetic or microbiological inventions. The answer seems to be plainly yes, of course. And, by the way, the question is not whether it's possible because it's real and everything that's real must by definition be possible. Therefore, we should concentrate on the question whether from the already patented and maybe future patentable procedures, tools, of course, and objects, some problems with the application of these patented things arise which are ethically dubitable. I think there are lots of open questions about what we should do and what we should not do with our genetic or microbiological findings.

Sahai: Vis à vis the question of patents per se. I don't think I would restrict it, I'm leaving it open. But the application of patents looks completely different with respect to the third world. This would, I think, come into conflict with the first and second categories that you raised about universal principles and regional principles, but I think very particularly on the regional principle of sustainable development which would be violated by the extension of patent or the application of the patent into areas that I have demarcated before.

Zimmerli: I am very glad you are raising this question, as I said before when we discussed this during the coffee break. I actually wanted to discuss this with you because I think what I said might be a little bit misunderstandable, because my task was not to say that there are no possible negative consequences of patenting with respect to the Third World and sustainable development. But my question was just we should not discuss this with respect to the question whether we should patent. So, if you ask me about the question of possible consequences of patenting, then of course we should enter the discussion of the question whether not only sustainable development but also the principle of fairness with respect to the Third World is violated by exporting special kinds of crops and seeds, for instance. Whether they are patented or not. Large companies will export them if they are patented or not. Therefore, this should be discussed under a different heading and I totally agree with you that there might be some additional advantages of a patented crop compared with a non-patented crop or seed which is genetically altered and which is better or has more advantages than other crops and seeds. But again I think this is not the question of the ethically viable patentability or not.

Sahai: The universal principle of fairness and justice I didn't think was being violated so much by sending the "wrong" kind of seeds. I was thinking more of the violation of principles of fairness and justice by the utilization of resources in perhaps not the most equitable way which would definitely come into conflict with the very fundamental question of using those resources for patenting.

Zimmerli: I just wanted to say that I agree, but then you said "for patenting" and there again do not agree. Of course, there are many violations of these principles. You have to run down the whole four level model with each question and my suggestion is not to overload the question of patenting with all the other much more important questions. And I would furthermore suggest that we should continue our talk on the other questions which are in my opinion vital questions.

Folman: What you are actually saying is that the realm of patents is not the right place to discuss various moral issues with regard to genetical engineering, molecular biology, etc. And then the question is what is the right forum to discuss those things? You may say politicians really represent the public in a way. The thing is that people in the Western world do not trust politicians very much. They trust the legal system much more, and the patent system is regarded as part of the legal system. When a patent is granted the public sees it as some sort of approval by the legal system of the use of genes. The public does not exactly understand what genetic engineering is and is worried, it is afraid of misuse of genes, of monsters, of manipulation, of various other things. So, at present the public does not get a good answer. The patent office says, that's not my problem at all, and maybe they're absolutely right. However, who is actually the body which should really discuss the public anxiety of manipulating genes. I feel that the legal system has a responsibility, maybe not a formal one, but it has a deep public responsibility to answer the public for various things it is doing because it can lead to many things which we do not yet know. I'm really afraid that the politicians may not be the right body to create the right laws to protect the public at least from the anxiety it lives in.

Zimmerli: I completely agree with what you're saying but again, let me just answer your question. Which is the body to discuss it? This conference, for example, like almost every conference of interdisciplinary participants, because this is a question we should discuss between scientists and lawyers and philosophers and politicians. Let us discuss these questions with respect to those legal regulations where they belong and this is not in patent law. But you're completely right with your first remark that there are two sides to it. The one side is where is actually belongs and the other side is what people think and how they feel and what they think of the legal system and what they think of the politicians a.s.o. In one word, the question of acceptance and acceptability of what we are trying to do in science and technology. And that again is a very interesting question. I gave an answer and I'm very glad to be able to repeat it. We should constitute (and are already doing so) committees and groups and research centers – risk discourse was the term I mentioned –, risk communication between policy on the one hand and, science on the other hand, the actual working politicians, lawyers etc. and the churches and the other groups within society. It's not easy but it's possible to do that – actually German legislation has done that to a great extent with the Enquête Commissions. They have discussed quite intensively, more intensely than any other nation I know of, the question of ethical implications, not of patenting, but of research with respect to the human genome, the question of reproductive medicine and the question of production of plants and animals. These are the different bodies in which these discussions should be put forward and I would say again I gladly agree to continue this discussion in different other groups but not on the patenting question.

Straus: You mentioned the politicians and the Enquête Commission. Just to show you how schizophrenic the politicians can be: they have at the same time recommended that many, many, things in this area should be supported by public funds and at the same time they said that they shouldn't be patented or they at least expressed their great hesitation to do that. Now, we are in a market economy, I'm not talking about India or the former Soviet Union. If you accept that the patent system offers very efficient means for incenting research and development in this area and at the same time you say, we will spend our money, but you shouldn't patent the results of the work financed publicly or privately, this is a contradiction and somewhat schizophrenic.

Zimmerli: Being Swiss I will not comment on foreign politics but I think it's not just the fault of the politicians, I think it's our fault, we are always blaming the politicians instead of looking at possible alternatives, and I can say once more that there are possible alternatives, for instance in discussion groups which are informal but which are opinion makers consisting of people who are able to influence other people. Therefore, it's actually our task and you're doing a good job, Dr. Straus, in convincing other people.

Beier: I would like to bring in the opinion of the federal government expressed in a statement of the Federal Ministry of Justice. I don't know if that statement has been made by politicians, but this is the opinion of the federal government and I would ask also Mr. Zimmerli if he can agree with it. I may be allowed to cite it in German. "Nach Auffassung der Bundesregierung wird in der Diskussion über die ethischen Fragen der Biotechnology nicht immer hinreichend präzise unterschieden zwischen der Sittenwidrigkeit und der öffentlichen Ordnung, moralischen Aspekten, einfachen gesetzlichen Vorschriften, die auf ethischen oder moralischen Erwägungen beruhen und eigentlichen ethischen Erwägungen. Nach Auffassung der Bundesregierung ist das Patentrecht nicht der geeignete Ort, um im einzelnen die Zulässigkeit von genetischen Veränderungen von Menschen oder Tieren zu regeln. Hierzu bedarf es zunächst eines gesellschaftlichen Dialogs, sodann entsprechender einfacher gesetzlicher Regelungen, insbesondere im Bereich der Humangenetik. Nachdem durch das Embryonenschutzgesetzverfahren die Veränderungen der genetischen Identität eines Embryos, Fötus oder Menschen einem generellen Verbot unterzogen worden sind, werden auch für die somatische Gentherapie und die Nutzung der Genomanalyse ergänzende gesetzliche Regelungen diskutiert und zum Teil schon vorbereitet. Diese Vorgehensweise stellt nach Ansicht der Bundesregierung den geeigneten Weg dar, um im Bereich der Biotechnologie und der Biomedizin ethische Aspekte einzubringen. Die Erarbeitung einer globalen Bioethik im Bereich des Patentrechts ist nicht möglich. Ethische und gesellschaftspolitische Zielvorstellungen sollten danach zunächst im Rahmen der allgemeinen öffentlich-rechtlichen Gesetzgebung berücksichtigt werden. Das Patentrecht kann und sollte solche Vorstellungen und Überlegungen jedoch dort berücksichtigen, wo genetische Veränderungen der menschlichen und tierischen Erbinformation die Grenzen der guten Sitten und der öffentlichen Ordnung überschreiten."

Hess: I think I should say just a few sentences about what has been read in German. The essence is simply that the government or the governing minister suggests that all these issues which you brought forward should be discussed in the public; public dialogue should be initiated and it is going on all the time. But that does not belong to the category of patent law which has a different type of function. It does belong to the category of what is legal and what is not legal in handling biological material of the living.

Zimmerli: I didn't know this statement otherwise I wouldn't have given my talk this morning but just read the statement. I think that with respect not only to ethical considerations what we understand by politics is in the process of changing considerably so that we probably will have some very important changes outside of parliament and outside of official party politics. And exactly that is what the German government itself is proposing, and I think that's a sign of prudence.

Grunwald: We have a problem with legislation on this kind of subject. I think what you, Mr. Zimmerli, told us this morning is correct: That professional standards are a good way to deal with these things. Now the declarations of Helsinki and Tokyo are, from this point of view, sufficient. We don't think it's necessary to go into legislation on this kind of subject, but if the legislator thinks so, and we learn that he does, he will go ahead. My point is the problems we discussed in these couple of days are so important that we should be unanimous before they start their legislation.

Hess: I think we had discussed quite a number of issues with respect to how to proceed in the political debate with respect to our contact to legislation. Every country has a different system but in general I think it works the same way. I think we should now focus a little bit more on other issues which came up during this meeting.

Berman: Being a Judge for 25 years I have been confronted, not once, with ethical issues and questions of morality and public order. Just to give one example, I can draw attention to the provision in contract law, which prohibits enforcement of a contract which is contrary to public order. In such cases it is for the judge to decide, on the merits of each case. There are neither legal provisions nor guidelines for defining *ethical* and *moral* issues. Nevertheless, the judge has to decide whether the performance of the contract, if enforced, will encroach on public order or morality. The Judge can't refer these questions to the legislator. The public relies on the judiciary, semi-judiciary or other discretionary bodies to be the quardians of such issues and principles. Mr. St. Rémy commented, yesterday, on the subject of pornography. And my question is "What is Pornography or Obscenity"? and "Who is to judge and decide on such issues"? To me, these questions are not different from the questions: "What are Public order and Morality"? and "Who is to judge and decide on it"? Judges are required to decide on all questions that legitimately arise in court during proceedings. It is true that *public order, ethics* and *morality* are not clear-cut issues but so are, also pornography and obscenity. A Judge is not expected to say that such issues are unclear and since the legislator has not defined such terms, the court may not pass decision. In the contrary, a judge must decide on all issues at dispute. So what did U.S. Judges decide on the question: "What is Pornography"? – "Pornography" is what the Judge says: – "It is Pornography"! The same approach may be taken by courts on questions of endangering the public order and morality by biotechnological inventions!! The courts are the first to be approached with new and "innovative" questions. The pace of science and technology is much faster than the pace of legislation and judicial competence. Considering the complexity of the new technologies and the unlimited possibilities of manipulation in the field of biotechnology, I personally would not be satisfied in leaving the question of encroachment on public order and morality by a biotechnological invention to be answered in the manner as was done in matters of pornography. I take the liberty of saying that perhaps biotechnoogy, as such, has outgrown the competence of the courts to understand what exactly is this new-old technology and mybe it is beyond the judge's ability to give meaning to the prerogatives of "Public order and Morality" in relation to biotechnological products or inventions. Therefore, I strongly suggest that such interdisciplinary meetings should enhance comprehensive explanations and guidelines for the judiciary ant the discretionary bodies in understanding the processes of genetic engineering and biotechnological inventions. But I also repeat, what I already said yesterday, that I believe that the provision *prohibiting* issuance of a patent which is against public order or morality is an important provision and should thus be retained in Article 53 (a) EPC. These provisions can be efficiently enforced and decided upon, during the process of examining and invention and before granting a patent. One should aim to prevent a possible disaster before its occurrence.

Hess: I don't want to go into the details of these problems but the committees of the G5 had suggested several times that the speed of research and new knowledge is such that the public will have problems in picking up the necessary knowledge. This poses special educational problems at all levels. Sources of education should be generated and made available to the public. Some countries like France really have amazing success in this respect, but not all countries have these facilities, and I think this is a very important point and I'm glad that you brought it up to keep in mind that this is also education of judges is necessary. I think that we should now come to some other points of view which, in my opinion, should be discussed in more depth. One is the problem which will come up in the future to quite an extent – what is the definition of a gene? It has been touched upon by Konrad Beyreuther already. What is a gene? One gene expresses one protein perhaps with two different functions but other DNA sequences, promotors or control genes are necessary for activation of up to hundreds of other genes in the course of early development. We know of master genes which activate at least 20 or 30 target genes. In this case patenting must give a precise definition of what a gene is and it would be very useful if Konrad Beyreuther would say a few words to this in addition to what he had already said. I would be very glad if you also would comment on the problem of copyright because that problem has not been brought up yet and I found it very interesting.

Zimmerli: There has been a problem with engineers. For years I did advise about 15 engineers regarding their code of ethics in the VDI, the German Engineer' Association. I would rather say we should not do that because it takes years to get guidelines and afterwards they are so abstract because they are the product of a consensus that you can't handle them. What I would suggest instead would be to invite two or three people from ethics and the field to each case and ask them for their expert opinion. In each case you will get a different picture because time changes the problems.

Beyreuther: What I think is lacking in politics and among lawyers is a kind of vision. We heard from Dr. Rowland that in the United States we had the Asilomar conference. The question was: Why shouldn't we stop this research in biotechnology? They didn't do it and also didn't start any legislation. We in Germany have this fascinating thing going on – the public seems to accept gene therapy which is based upon gene technology which they don't accept. This is not schizophrenia, this has something to do with feelings, some people feel pregnant, others not, some feel they are shortly before fatherhood – they never become fathers – there must be something in neurobiology that makes those people have these feelings. They move on irrationally and say, let's do gene therapy without having basic knowledge of it. I would like to make a comment to what we as basic scientists feel are the genetic resources that we are distributing now. I really don't have serious concerns that anything is going wrong because Dr. Yorke did a fantastic job by throwing into the discussion this issue of cross-licensing. I'd like to address the issue once more that is my major concern. The number of genes is limited that we're dealing with and we are going to sell those genes of commercial interest right now. Those genes are defined by the product and we still have this absolute product protection. This makes sense somehow on the basis of the decision that has been derived in the United States where 400 or so companies involved in bioltechnology have more or less agreed to claiming genes only if the complete coding region and its biological functions are known or more precisely they ask for exclusive licensing only of substantially full-length cDNA. I think we can live with that provided that the cross-licensing principle is maintained to the generous extent it was in the past. However, there remains some concern regarding these resources. In the past, the number of potential pharmaceuticals is hard to give, maybe several million potential drugs that chemists can produce, maybe billions, I don't know. The number of genes that can be patented is very, very, limited. You have genes that are used by all cells, so-called "house-keeping" genes, 1000–2000 genes allow a cell to live and to reproduce. And I think it's too late to discuss whether those genes should be patented or not, they are patented and they deserve absolute product protection. But it's not an issue of whether this gene is a human gene or not a human gene, the issue is of resources and we'll have to live with this issue for 12 years. The resources are in part distributed an we are going to distribute them in the next years as soon as the 3000 disease genes that are left are discovered and are patented.

Hess: What about the non-"house-keeping" genes?

Beyreuther: This is a very, very, touchy thing. I don't think there's any danger for mankind within these genes but they are discovered by the parmaceutical industry right now through our help. We said that the future drug will address regulation and those genes will also be distributed and will also fall under this regulation. It will be very important that we continue with the practice of cross licensing. I work with a gene that has probably a dozen binding sites and even a world company can only assess one binding site in regard to drug development and if they don't cross license the gene for other development, I think that would not be responsible, but I'm not really concerned if we continue to go on in the way we did it in the past.

Izhar: I think we are addressing the problem of patenting single genes or sequences. I heard from Barbara McClintock that the most significant single genetic contribution to human welfare was the invention of F_1-hybrid in corn at the beginning of the century in the second decade. The most important traits are probably the quantitative traits which have to do with food production whether it's in plants, animals etc. You can compare it to the genes that we are talking about now, mainly for medical purposes. What about these?

Beyreuther: You now addressed a very important question: What is the difference between primates and human beings and what is the difference between less and more? I think one can foresee that the major difference will be a quantitative one, not a qualitative one. If you talk about these genes that Dr. Hess threw into the discussion, these regulatory genes, these are the genes behind which there is a lot of money. They also are very important in determining a species. But they are found everywhere and in addition to the genes you mentioned are the "jumping" genes which are also very important and they may disturb everything we are talking about. It is the interaction of genes that is important.

Hofschneider: I think it's really very important that we define very precisely what a gene is with respect to patent rights which can be granted. A gene is much more than a base sequence defining an open reading frame. At the same time it can have completely different properties which you cannot get out by only having the sequence. It could be an enhancer, it could be a spacer in the genome. There are many possibilities. A promoter may have very important regulatory functions which cannot be defined by only having the sequence. So, it would be a mistake if somebody who has only the sequence or only reading frames gets all the rights for things which he has not detected. I think one should be very, very careful and only if you have a precise definition of a gene can we distribute rights of intellectual properties.

Hess asks Hofschneider as a panel member to say something about the interaction between basic science and industry in terms of patents.

Hofschneider: I don't want to discuss it in detail, I think it would become repetitive. The only thing which one may add is that the interaction between industry and basic science in Germany could be better and probably many intellectual properties of this country get lost not because we are going for too many patents but since we are going for too few. I think it would be important to educate the scientists how to go for a patent. In the Max Planck Society we have "Garching Instruments" and so we get some advice on how to do it and in the so-called Grossforschungseinrichtungen like the German Cancer Research Center the situation is similar, but our poor colleagues from the universities do not get any advice. At the same time they have the "German disease" as Prof. Zimmerli defines it. That means it is regarded as unethical if you want to make a profit from something which may be patentable. That is a very bad situation, but it's a national problem. At least the German colleagues who are here now from both sides – from patenting offices as well as science – should do their best to improve the situation here in Germany.

Garthoff: Maybe I will comment on that last issue – the question that was also raised in context with regard to cross-licensing. It must be in the interest also of industry to exploit that one invention that was made to the very last point. Maybe some other inventor comes up with a new indication, for example, some different kind of gene regulation that he hasn't thought of in the first place – with a different point of view. The new point of view that came up, let's say, from the Max-Planck-Gesellschaft, would then be patented as well and there would be a cross-licensing in agreement. This is done in industry almost every day. It would be done also between industry and university. As you said before, we rather have the problem in Germany that the German scientist at the university is too far away from the industry. The other issue I would like to approach is something that you put forward as a question to me before – what is a "middle-aged scientist" in industry in the future? Let me explain just a few things to give you some idea what people in industry might be concerned about. One issue was raised before that there might be no patents for transgenics, in this case not for research, but in addition for producing valuable therapeutics. If that would hold true in the future, and if we wouldn't have a chance to patent such animals, the industry might not be interested in investments anymore. The other thing is that by now everybody thinks in terms of moral and benefit assessments in relation to the European Patent Office. We absolutely object to that. I think that would be totally inadequate. The other thing I would also like to mention is the issue of world-wide differences the colleague from Sandoz just mentioned before, nothing is worse for industry than uncertainty and we had a lot of uncertainty specifically in this country with regard to gene technology in the past. If there is uncertainty, you won't invest. So, if there is world-wide uncertainty in terms of what can be patented, you have a problem. The issue of first invention in the U.S. creates another problem. You have to do something to overcome that. I would like to put forward another thing – the problem in the future. Every one of us is convinced there should be no handling and no research into the germ line. But what if – and that would be a question I would like to put forward to the colleagues from EPO – for example, one of the EC countries, for example, Spain, would come forward and say, in our country germ line research is permitted and a Spanish company would come to the European Patent Office and would ask for a patent on that. In essence, I think the European Patent Office will have to allow this patent even if it's not a valuable one in Germany, but it may be a valuable one in the rest of the world. And what do we do, if one country in the world comes up with a germ line issue or what if Craig Venter asks for fragments of sequences to be patented without knowing what the use of that is? That is the next thing I would like to address

– patenting of fragments without any use. I absolutely would object to this and it is totally different than in the past patenting practice.

Grunwald: Thank you Dr. Garthoff for that clarifying statement. I would like to ask you whether you would agree with Mr. Yorke that we could discuss, let's say, clarification – I would avoid the term extension – of the research privilege.

Garthoff: It's easy for me to say yes, sure, let's discuss. But I think you go a little further than that. Research exemption is one issue. I think that's an easy one because as Yorke mentioned, it's in the interest of industry to be in a good collaboration with your partner who is the research scientist on the bench at the Cancer Research Center or wherever. So it's in our own interest to do that. That's why everybody can use PCR without any problem. You don't have that kind of a written exemption but at least it's in fair agreement. I think that was put forward in the past as well. Let's rather go on fair agreements than to put a law into effect when we don't know what the future might bring. So, I think that's the basis of our direct collaboration and I think that we would always like to continue with that collaboration.

Hess: I wanted to make just one remark in terms of the German chronic disease. I think this is very much related to the field. It's no doubt that in certain fields, even the Max Planck Society has been extremely successful in applying for patents, for example, in chemistry and physics. In life sciences, it is different, but on the other hand we should not overestimate the number of people who have the problem. Just in the last 10 years or so, extremely important patents have been brought forward and applied.
Are there any questions to be addressed to Prof. Ganten from the medical side?

Straus: My question is linked to the lecture of Prof. Beier and touches upon the problem of medical methods. I was always very curious and never had the opportunity to ask high-ranking physicians, how is it really in the practice? How large is the time lag between the leading figures in the field and the followers? I know that many people have to go to a certain place for surgery, how long does it take to bridge the difference between Hannover with Prof. Pichelmayr as surgeon and other places in Germany, for instance?

Ganten: This cannot be answered in a general way. But definitely in some hospitals – the technical capabilities, also the intellectual level, are not state of the art. On the other hand, there are top hospitals that are on an international level, they get all the information, they read the journals. One of the reasons I would tend to say that some of the other clinicians may be angry with me and contradict – the general level of clinical medicine in anglophone countries tends to be better than in German speaking countries and I think the reason for this is that most, if not all, of the important original literature is being published in English and the general clinician does not read English. They wait for translation or for secondary reports so that continued education at the clinical level is going on in the German language, and that results in a certain lag time of promotion of results into practice. This is the information gap. Of course, this is half a year or a year. One of the reasons is that the education is not oriented towards original papers but towards textbooks which again lag behind, – German textbooks lag behind English textbooks. If we want to do something here we have to introduce, more than we do at the present time, the English language which is the language of science into the educational system for medical doctors. It is the working language in the laboratory, and in most university hospitals. But it's a traditional language problem rather than a problem of patents and that kind of thing which we are discussing here.

Straus: I may just make a nasty remark and ask the colleague from the EPO to bring the message to his president that the common conviction that all people in all countries of the European Community should read patents in foreign languages because everybody knows English appears doubtful. Listening to you and realizing that even physicians in this country have problems in reading English language literature, isn't it true?

Ganten: Everybody learns English, everybody can read English but they are not used to looking into international journals for their daily instructions.

Hess: Mr. Hofschneider and I have been involved for quite a large number of years in increasing the density of clinical research without too much success. I think that still we are lagging behind the English system which has much more possibilities for interaction between basic science and clinicians. It is better than it was 15 years ago, I agree, and your institute is supposed to be a good example for this type of interaction. But I see that the education of doctors in terms of applying these modern techniques is not as it should be and the reason for this is certainly that clinical research in molecular biology and genetics is not implanted in the clinical system as it should be, as it is perhaps to some extant at Freiburg University and here, of course, in the Cancer Research Center and in your place. I think this is one of the major reasons why application of all these techniques whether it's diagnostic or not is lagging behind. That's my interpretation.

Ganten: I certainly agree with that and there are some institutional barriers there in the traditional German university system which probably support this type of separation of basic research and clinical research and practice. I think, without elaborating on that, molecular biology and gene technology may make the difference. It is our feeling that young clinicians who have learned molecular biology during their studies, realize the techniques are much simpler. Before it was impossible to do sophisticated protein chemistry and at the same time do clinical practice – go to the patient's bed. Now, with molecular biology the techniques are really simple, they are standardized, there are kits, you buy them, most of them are patented, they're expensive, but you just take them from the shelf and within half a day or a day or two days a week you can really accomplish very precise, excellent, basic research and at the same time have some clinical practice and maintain your clinical knowledge. So, I think, if molecular biology really gets into the clinic, which it slowly does, this will change the picture.

Beier: I think one has to consider different points. In my opinion, what the Max-Planck Society did as a model system was very successful. Apparently not only Freiburg, but now Tübingen and Ulm have adopted this system. Secondly, I think there is one mistake if we compare the German medical care with top hospitals in the United States like Massachusetts General Hospital. Obviously the average here is not as as good as the Massachusetts General Hospital; however, if you compare the average of this country with respect to medical care with the average of the United States, I would say it's at least as good as in the United States. What would you prefer, to get medical care in a small village in Germany or somewhere in Arizona in the desert?

Hess: I was very glad that this point was brought up and I see we have some different views. I'd like to slowly come to a close and would like to suggest to listen to Prof. Beier, the Nestor in that field. He has three points he wants to read in terms of summing up what has to be done in the field of internationalization.

Beier: We were asked by the organizers to propose principles of an international understanding and we find that such an international understanding concerning the patentability of biological organisms and human genes should be developed along the following lines:

1. Patent exemptions for specific fields of technical innovation constitutes the ultima ratio of patent law and need compelling reasons to be justified. There is no such reason with respect to biological inventions including product inventions whether microorganisms, plants or animals. The human body and its parts as such per se should never become the object of patent protection and caution is necessary with respect to inventions that use the human body or its parts for producing valuable substances.

This is the first point. I think it is somewhat a resume of our discussion in these two days.

2. The protection of inventions or discoveries which are located at a rather early stage of the innovative process of research and development, research results which are closed to industrial application, has to be secured in principle in order to include the scientists in the Belohnungs- und Anspornsystem (reward and incent system) of the patent system. There is, however, a current tendency of applying too early and of searching for overprotection that may have a stifling effect. Possible remedies include a stricter application of the disclosure requirement, to restrict the scope of protection to the claims that include an actual obvious use of these products and maybe also a legislative introduction of dependency licenses if the cross-license system does not function.

3. Inventions should be protected in such a way that not only the results of industrial research, but also the results of publicly funded research will find their way to the market and by that lead to economic and social progress. The research process itself, should however, not unduly be restricted by patent rights. There is a good case for clarifying and possibly broadening the research exemptions which are already provided in most national patent laws and in the European patent law in order to promote a fruitful coexistence of patent law and academic research. I have the impression that this is also in line with that what we have discussed here. We are scientists, lawyers, people from industry and no politicians. It is now up to us to convince the people who have the say and to bring up our ideas to governments and to the legislative bodies.

Grunwald: I think what you said, Dr. Beier, was a fair summary of the discussion. I would like to add the point that we were all agreeing upon, the need of adding a grace period into the European Patent Convention and national law.

Gugerell: Prof. Beier, you say that the human body and the parts thereof per se should not be patentable. How do you interpret the per se?

Beier: It depends on the way of claiming. I'm not a patent attorney and I don't know how to claim correctly, but I think there is a difference between the protection of the human body as such and parts of it as such without any further element of invention. Maybe Mr. Straus or Mr. Moufang can give you a more correct answer.

Vogel: Ladies and gentlemen, we are at the end of this meeting. I think it was quite a good meeting. I also liked the discussions in this meeting. They were controversial but fair and I think this is always necessary. I must say, of course, we could not arrive at a full agreement on all the points we have discussed. The feeling that remains for me is a slight feeling of uneasiness that we are talking as if we were alone in the first world, but we should keep in mind all the third world problems that have been discussed here. I think this will be a continuous challenge for us also to think about those problems and to try to do our part for a reasonable solution. And I think that the discussions that were started here will certainly by carried on. It was intended as an interdisciplinary activity and I think this was successful as such.